# THE CENTER, BULGE, AND DISK OF THE MILKY WAY

# ASTROPHYSICS AND SPACE SCIENCE LIBRARY

A SERIES OF BOOKS ON THE RECENT DEVELOPMENTS
OF SPACE SCIENCE AND OF GENERAL GEOPHYSICS AND ASTROPHYSICS
PUBLISHED IN CONNECTION WITH THE JOURNAL
SPACE SCIENCE REVIEWS

VOLUME 180

# THE CENTER, BULGE, AND DISK OF THE MILKY WAY

Edited by

LEO BLITZ

*Laboratory for Millimeter-wave Astronomy,*
*University of Maryland, College Park, U.S.A.*

SPRINGER-SCIENCE+BUSINESS MEDIA, B.V.

Library of Congress Cataloging-in-Publication Data

The Center, bulge, and disk of the Milky Way / edited by Leo Blitz.
    p.   cm. -- (Astrophysics and space science library ; v. 180)
  Includes bibliographical references and index.
  ISBN 978-0-7923-1913-9         ISBN 978-94-011-2813-1 (eBook)
  DOI 10.1007/978-94-011-2813-1
  1. Milky Way.  2. Astrophysics.  I. Blitz, Leo.  II. Series.
  QB857.7.C39  1992
  523.1'13--dc20                          92-24067

---

*Printed on acid-free paper*

# TABLE OF CONTENTS

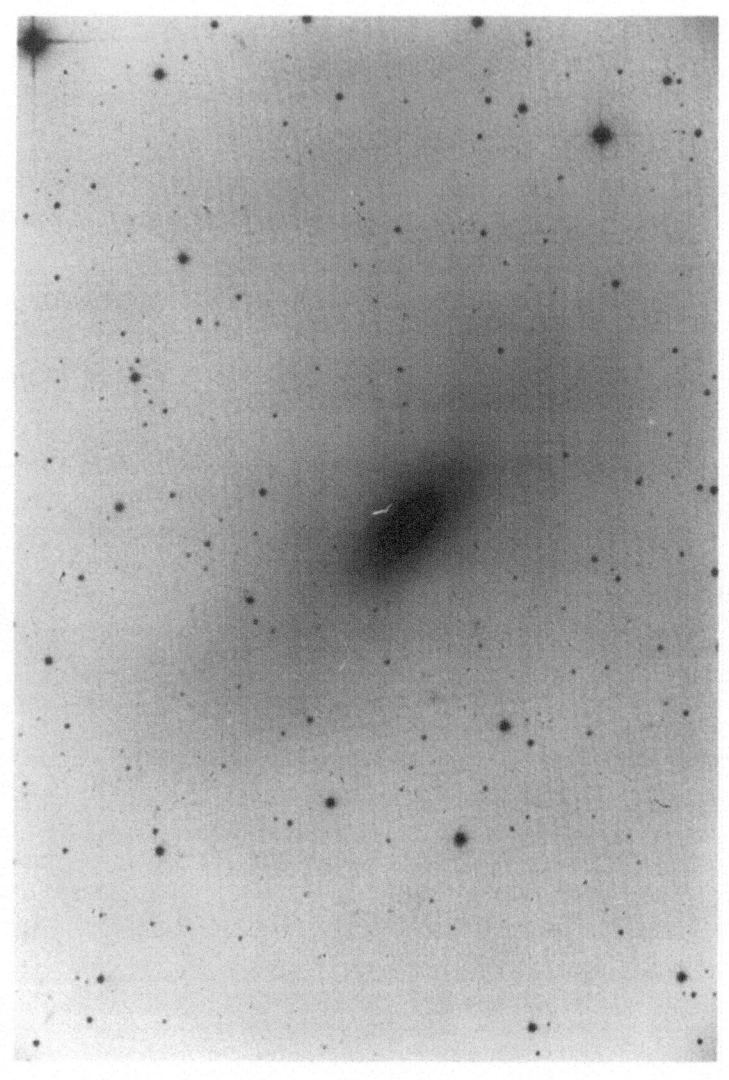

*Frontispiece:* NGC 5102 has the morphology of a normal S0 galaxy, but has a b
nucleus with the spectrum of an A0 star (see Figure 17). The entire bulge of this gal:
has Balmer lines, so we may conclude that it has been involved in the same starbu
that formed the nucleus ≈ 1 Gyr ago. We are reminded that bulges can form late in
life of a galaxy. Plate (103aO+GG13) obtained with du Pont 2.5-m telescope at the l
Campanas Observatory; North at top, East to the right.

# PREFACE

This little book is the outgrowth of three sessions of invited papers given at the IAU Genaral Assembly in Buenos Aires in 1991. Normally, the papers given at these sessions are not written up, but what we found was that the authors were moving the conceptions of the Galaxy in some new directions and we all agreed that it would be a good thing to write the results up and gather them together in a single volume. The articles are of uniformly high quality, and much of what is presented here is new, not just review. Gezari's stunning high resolution images of the Galactic center made with his 10 $\mu m$ infrared array camera are presented here for the first time.

The volume is organized in the conventional way, starting from the center and working out in order of increasing distance. The articles by Yusef-Zadeh and Wardle, and by Gezari on the inner parsec of the Galaxy present in the first instance a synthetic view of the diverse phenomena at the center and in the second, outstanding images from a newly opened array window. Rich's beautifully argued paper presents a strong case for a bulge which formed subsequent to the globular cluster population. The Spergel paper is an attempt to unify work he and I have done arguing for the presence of a bar and also for a larger triaxial spheroid that controls the kinematics of the disk. The article by Whitelock is a new analysis of the IRAS sources at the Galactic Center arguing that the Milky Way has a barred spiral structure. Liszt's paper presents a new look at an old subject: the distribution of HI in the disk of the Galaxy, and Bronfman's article is an outstanding synthesis of molecular line surveys of the Galactic plane, but also contains new results from high density CS surveys. Latham's article presents the most recent information from population studies about trying to identify whether the thick and the thin disks are distinct stellar populations.

All in all, those who attended the sessions were rewarded with an unconventional program with outstanding new results about the large scale structure of the Milky Way. This volume is an attempt to share those results with a larger audience.

Leo Blitz
College Park
May 1992

# LIST OF ADDRESSES

Leo Blitz,
> *Laboratory for Millimeter-wave Astronomy, Department of Astronomy,*
> *University of Maryland, College Park, MD 20742, USA*

Leonardo Bronfman,
> *Universidad de Chile, Observatorio Astronomico Nacional,*
> *Casilla 36-D, Santiago, Chile*

Robin Catchpole,
> *Royal Greenwich Observatory, Madingley Road,*
> *Cambridge CB3 0EZ, United Kingdom*

Dan Gezari,
> *NASA Goddard Space Flight Center, Laboratory for Astronomy/Solar Physics,*
> *Code 685, Greenbelt, MD 20771, USA*

David Latham,
> *Center for Astrophysics, 60 Garden Street, Cambridge, MA 02138, USA*

Harvey Liszt,
> *National Radio Astronomy Observatory, Edgemont Road,*
> *Charlottesville, VA 22903, USA*

Michael Rich,
> *Department of Astronomy, Columbia University, Box 52 Pupin Hall,*
> *538 West 120th Street, New York, NY 10027, USA*

David Spergel,
> *Princeton University Observatory, Peyton Hall, Princeton, NJ 08544, USA*

Mark Wardle,
> *Northwestern University, Department of Physics and Astronomy,*
> *Evanston, IL 60201, USA*

Patricia Whitelock,
> *South African Astronomical Observatory, PO Box 9,*
> *Observatory, Cape 7935, South Africa*

Farhad Yusef-Zadeh,
> *Northwestern University, Department of Physics and Astronomy,*
> *Evanston, IL 60201, USA*

# A COHERENT PICTURE OF THE INNERMOST PARSEC
# OF THE GALAXY

FARHAD YUSEF-ZADEH and MARK WARDLE

*Department of Physics and Astronomy, Northwestern University*

**Abstract.** This review attempts to provide a framework within which recent radio and infrared results related to the inner parsec of the Galaxy can be interpreted. The unifying theme is the outflow from the Galactic center. This wind arises predominantly from the IRS 16 complex which lies within $1''$ of a massive black hole candidate. We present compelling morphological and, in some cases, kinematic evidence for the interaction of this outflow i) with a mass-losing supergiant, known as IRS 7, located within the inner light year of the dynamical center of the Galaxy, ii) with the continuous flow of ionized gas associated with Sgr A West which is orbiting the Galactic center, iii) with the massive black hole candidate which is thought to be coincident with a nonthermal compact radio source, known as Sgr A*, and iv) with the circumnuclear molecular disk which engulfs the inner few parsecs of the Galactic center.

## 1. Introduction

The Galactic Center is obscured at visible wavelengths by more than 28 magnitudes of intervening dust and gas. Most of our understanding of this region is, therefore, based on observations in the infrared or at longer wavelengths (see reviews by Brown and Liszt 1984 and Genzel and Townes 1987). This strong radio and infrared source is often the first object examined with new instrumentation or with new techniques. Indeed, it was present at the birth of radio astronomy when Karl Jansky (1932) first reported the detection of radio emission from the Galactic center. More recently, advances in radio and infrared imaging techniques have revealed an enormous morphological complexity in this region, and numerous components with different scale sizes have been recognized. Because of this complexity, a consistent understanding of the relationship between different features is lacking. In this review we shall concentrate on the activity in the innermost parsec, and try and present a coherent picture of what is happening there. We begin by outlining the morphology of the region.

Sgr A* is the brightest radio source within the inner few degrees of the Galactic center region, and is located very close to the dynamical center of the Galaxy. An unusual bright, early-type infrared complex, IRS 16, lies within $\sim 1''$ of Sgr A*, as drawn schematically in Figure 1. IRS 16 consists of at least 15 components at $2\mu m$ (Eckart *et al.* 1991), and appears to be the source of a strong wind with velocity of order $700 \, \mathrm{km \, s^{-1}}$ (Hall, Kleinmann, & Scoville 1982; Geballe *et al.* 1984, 1987, 1991; Allen *et al.* 1990). Both Sgr A* and IRS 16 lie within an ionized cavity at the center of a ring of neutral material (the circumnuclear "disk", see reviews by Genzel 1989 and

1

*L. Blitz (ed.), The Center, Bulge, and Disk of the Milky Way, 1–22.*
© 1992 *Kluwer Academic Publishers.*

Genzel and Townes 1987). Within the cavity, three "arms" of ionized gas are in orbital motion around Sgr A*/IRS 16 (Ekers *et al.* 1983). The closest segment of ionized gas is known as the Bar within which a "mini-cavity" has also been identified (Yusef-Zadeh, Morris and Ekers 1989, 1990). The kinematics of the ionized gas suggests a continuous flow along the individual arms, some aspect of which is consistent with orbital motion of gas around Sgr A* or IRS 16 (Lacy *et al.* 1980; Serabyn *et al.* 1988; van Gorkom *et al.* 1984).

In the next two sections we consider the two objects, Sgr A* and IRS 16, which dominate the energetics of the inner parsec. Sgr A* is the best candidate for the concentrated mass which drives the dynamics of the inner parsec, whereas IRS 16 is the source of the ionized outflow from the region. We then discuss the evidence for the interaction between these sources. In particular, we suggest that the luminosity of Sgr A* is a result of the capture of a portion of the outflow from IRS 16. We go on to explore the dynamical consequences of the outflow in order to explain the morphology of the orbiting material in Sgr A West.

## 2. Sgr A*: A Million Solar Mass Black Hole?

Sgr A* is the brightest radio source within the inner few degrees of the Galactic center region, and is located very close (if not coincident) to the dynamical center of the Galaxy, embedded within an evolved stellar cluster. It is compact having a size < 20 AU and based on VLBI measurements it is elongated at 8.4 GHz along a position angle of $82^0$ with an axial ratio of 0.53. It is variable, and shows nonthermal characteristics with a spectrum resembling the cores of extragalactic radio sources (Lo 1989; Zhao *et al.* 1989; Jauncey *et al.* 1989). These sources are thought to be powered by accretion onto a massive black hole, thus it is not surprising that Sgr A* itself has become a black-hole candidate. Indeed, it is suspected that the nuclei of most, if not all, galaxies contain a massive black hole in a quiescent state. The behavior of the velocity dispersion of the stars in the nuclei M31, M32, NGC 4592 and NGC 4594 are consistent with the presence of a black hole (Dressler and Richstone 1988; Kormendy 1988). Perhaps the best evidence for massive black hole is seen toward the center of M31 with a mass range $4 - 5 \times 10^7$ M$_\odot$. As such, this estimate is more than an order of magnitude higher than the mass estimate for Sgr A*. Curiously, if Sgr A* were placed at the distance of M31, the surface brightness of the compact radio source associated with M31 would be weaker than Sgr A* by more than an order of magnitude (P. Crane, private comm.)

A large mass for Sgr A*, of order $3 \times 10^6$ M$_\odot$, is indeed suggested by its low proper motion (Backer & Sramek 1982), by the velocities of the ionized gas orbiting the Galactic center (Serabyn *et al.* 1988; Lacy, Achtermann

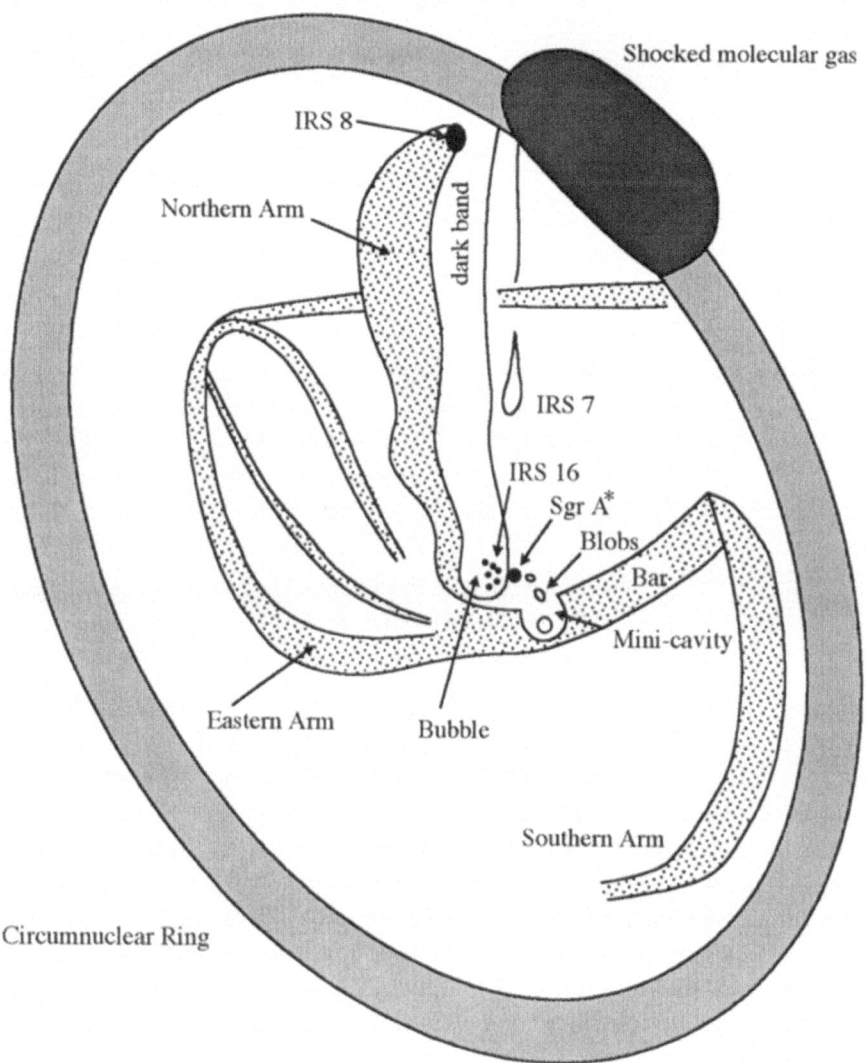

Fig. 1. A schematic diagram showing prominent features of the inner two parsecs of the Galaxy.

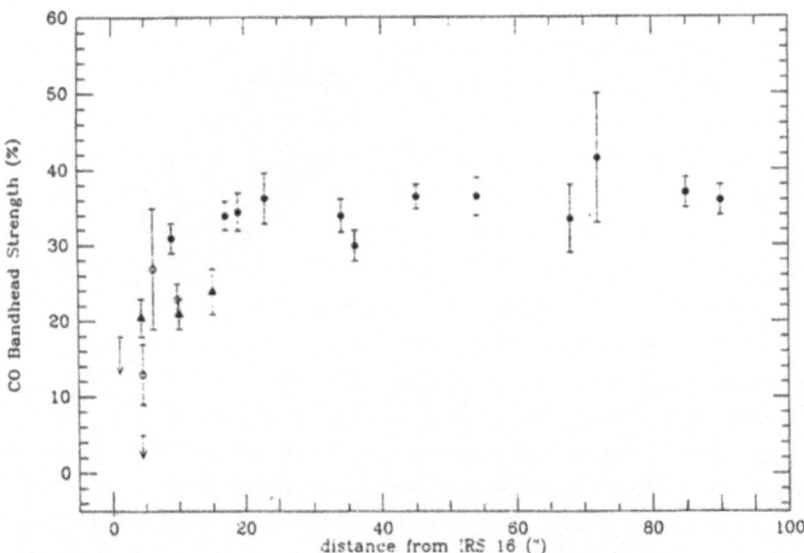

Fig. 2. Strength of the 2.3 $\mu$m absorption CO bandhead as a function of distance from the IRS 16 complex (Sellgren *et al.* 1990).

& Serabyn 1991) and of the late-type stars in the cluster that engulfs Sgr A* (McGinn *et al.* 1989; Sellgren *et al.* 1990). The inferred mass need not reside in the form of a black hole, however, because the material may lie anywhere within the central 0.5 parsec of the cluster, possibly as a compact star cluster or a population of faint, low mass stars (Sellgren *et al.* 1990). McGinn *et al.* and Sellgren *et al.* used the integrated 2.3$\mu$m CO bandhead to determine the kinematics of the cluster, but the strength of the bandhead declines in stars lying within 0.6 parsec of the center. This dramatic feature of the stellar emission, illustrated in Figure 2, could be produced by the local radiation field or by collision of giants with other stars (Sellgren *et al.* 1989). Alternatively, it could result from interactions with an ultra-high velocity wind from IRS 16/Sgr A* (Yusef-Zadeh and Morris 1991) or with numerous high-velocity outflows recently discovered by Krabbe *et al.* (1991). In any case, it will be difficult to obtain kinematic information from the stellar component on scales much smaller than the core radius $r_c \approx 10''$ because most of the stars along the line of sight lie at physical distances from Sgr A* that are much larger than the projected distance.

An alternative approach (Wardle and Yusef-Zadeh 1991) is to look for signatures of gravitational lensing by a million solar mass black hole. The characteristic angular scale of these effects ranges from ten to several hundred milliarcseconds, requiring the mass to lie within a hundred to a thousand AU of Sgr A*. Wardle and Yusef-Zadeh examine the manifestations of lensing i) on the appearance of the stellar cluster with $r^{-1.8}$ density distribution, ii) on a recent bright transient radio source which is located about 30″ southwest of Sgr A*, and iii) on Sgr A East, the nonthermal radio source lying behind but within a kiloparsec of the Galactic center (Yusef-Zadeh and Morris 1987; Pedlar et al. 1989). The most important effect in the case of the stellar cluster is the strong amplification of a few of the numerous faint stars above a given limiting magnitude. Strong amplification requires a high degree of alignment with Sgr A* so is transient on a timescale of months to years because of stellar proper motion. In practical terms these observations require 10 milliarcsecond resolution at $2\mu m$, which could be obtained by interferometry with a baseline of 40 meters. A further avenue for investigation is the lensing of stars in the Galactic bulge which tend to lie at kiloparsec distances from Sgr A*, with correspondingly larger Einstein rings. Another possibility to probe the nature of Sgr A* is by lensing of the Galactic center transient source which appeared recently in projection $\approx 30″$ from the the position of Sgr A* (Zhao et al. 1991). The intrinsic flux density of this compact source changed continuously over a month and became comparable to Sgr A* in its surface brightness (Roberts, private comm.). This new source could, in principle, change the flux density of Sgr A*. This situation could occur if the transient source were an extragalactic radio source or about 1 kpc behind the Galactic center. Although the change in the flux density of Sgr A* resulting from the lensing of the faint, secondary image of the transient source would be about 1 mJy, the change would be in phase with the increase in the flux density of the transient source. Finally, lensing of Sgr A East is of more immediate interest because of the higher resolution obtainable at radio wavelengths. Substructures associated with Sgr A East are recognized to lie at a projected distance of 10″ away from the Galactic center (Pedlar et al. 1989), so a secondary image should lie at $\approx 4(r/100\,pc)^{1/2}$ milliarcsecond. Surface brightness is conserved by lensing, so that given the morphology of Sgr A East, one may predict the appearance of the secondary image. Indeed, there is already evidence for extended structure near Sgr A* on a 20 milliarcsecond scale at $\lambda 6cm$ (R.L. Brown, private comm.).

## 3. The Wind from IRS 16

Another unusual bright infrared source—known as IRS 16—lying within $\sim 1″$ of Sgr A* has occasionally been cited as the key energy source at the Galactic center. IRS 16 consists of a number of infrared components, all of which have blue colors relative to luminous evolved stars. Allen et al. (1990) have pointed out that the line emission arising from IRS 16 subcomponents has a spectrum which is consistent with that of mass-losing WN stars. There are two lines of evidence for the presence of an outflow from IRS 16. The detection of broad He I, Brα, and Brγ emission lines from the vicinity of

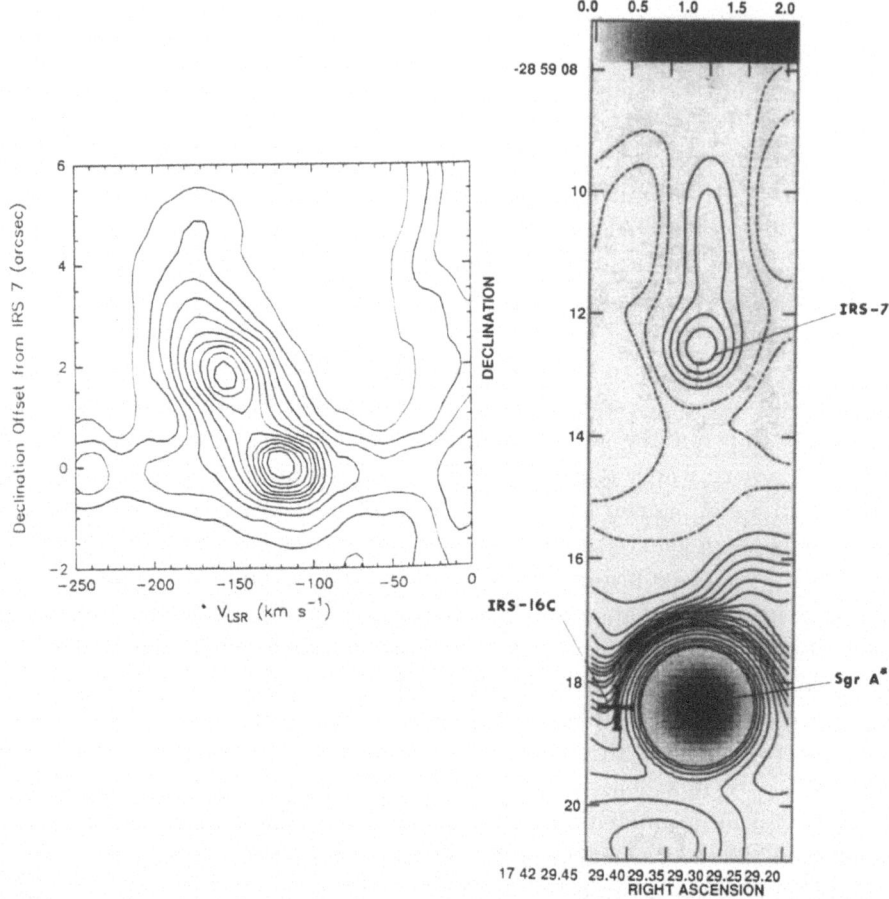

Fig. 3. The figure to the left shows declination-velocity diagram of 12.8 μm Ne⁺
emission from the ionized tail of IRS 7 (Serabyn, Lacy and Achtermann 1991). The
figure to the right shows the location of IRS 7 and its ionized tail, as seen in radio
continuum, relative to Sgr A* and IRS 16C (Yusef-Zadeh and Morris 1990).

IRS 16 indicate an outflow with a terminal velocity of 500-700 km s⁻¹ (Hall,
Kleinmann, & Scoville 1982; Geballe *et al.* 1984, 1987, 1991; Allen *et al.*
1990). In addition, the detection of $H_2$ line emission from the inner edge of
the circumnuclear ring that surrounds Sgr A West could be due to molecular
gas being shocked by the outflow (Gatley *et al.* 1986).

The case for such a wind has been strengthened recently by detection of
ionized gas from IRS 7, a supergiant within the projected distance of one
light year (Rieke and Rieke 1989; Yusef-Zadeh and Morris 1991; Serabyn,
Lacy and Achtermann 1991). Radio continuum observations have revealed

a cometary "tail" of ionized gas from IRS 7 directed away from the dynamical center of the Galaxy (Yusef-Zadeh and Morris 1991; Serabyn, Lacy and Achtermann 1991). Figure 3 shows radio continuum as well the kinematics of ionized neon emission from IRS 7 and its tail which extends in the direction northward of the Galactic center. The velocity structure of the tail shows clearly a continuous flow of ionized gas with increasing velocity from $-120 \, \mathrm{km \, s^{-1}}$ to $-180 \, \mathrm{km \, s^{-1}}$ in the direction away from the Galactic center (Serabyn, Lacy and Achtermann 1991). The tail is therefore definitely associated with the supergiant star. The cometary tail is interpreted as an initially spherical wind from IRS 7 that has been swept back either by radiation pressure, the ram pressure associated with the circumnuclear wind, or by the drag as IRS 7 moves through a stationary ambient medium. Detailed modeling of the shape of the bow shock seen in the ionized emission favors the circumnuclear wind hypothesis (Yusef-Zadeh and Melia 1991). Superimposed on higher resolution radio map in Figure 4 is the simulation of the surface of the bow drawn to the same scale, for four different values of the parameter $F$, which measures the relative strengths of the Galactic and stellar winds. The required location of IRS 7 is consistent with that due to near-infrared astrometric measurements as shown by a cross (Becklin *et al.* 1987). The partial flaring of the shock surface arises partly because the Galactic wind originates from a point (a projected distance of one light year distance away from IRS 7), and would be markedly reduced if it were due to the drag of IRS 7 through a static medium at the Galactic center. The curve labelled 3 in Figure 4 giving the best fit ($F \approx 50$) places the position of the central wind source within the complex of infrared sources IRS 16. We note that while the apex of the bow shock appears to be pointed toward the IRS 16 complex whose relative position is shown schematically in Figure 2, the tail of ionized gas associated with IRS 7 appears to be directed toward Sgr A* (see below).

## 4. The Interaction with Sgr A*

There is growing morphological evidence which will be shown in the following two sections, for the outflow and its interaction with Sgr A* and the Northern Arm. The interactions offer an opportunity to constrain the properties of the wind and to judge the dynamical importance of the outflow. Almost all components of IRS 16 lie to the east of Sgr A*. Figure 5 shows $\lambda 6$cm map of the inner $8''$ of the Galactic center where we note the most prominent components of IRS 16 are all distributed to the north and east of Sgr A*. We also note a bubble-like feature with a diameter of $\approx 3''$ forming around the IRS 16 complex as the western, southern and eastern edges of this bubble appear to terminate at the positions of Sgr A*, the Bar and the Northern Arm, respectively. Further evidence is suggested by radio continuum image

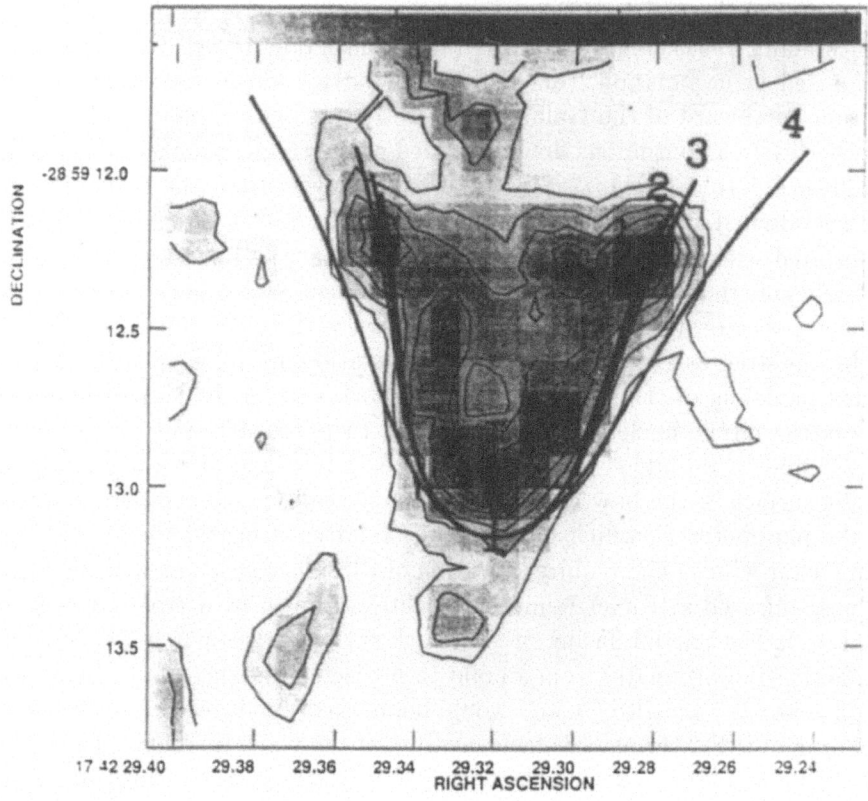

Fig. 4. Plot of outer shock radius for 4 different theoretical values of F, 500, 100, 50, 10 corresponding to plots 1, 2, 3, 4 superimposed on the bow-shock structure of IRS 7 (Yusef-Zadeh and Melia 1991).

of Figure 6 where a dark north-south band stretches from IRS 16/Sgr A*. It is quite plausible that such a structure is associated with the bubble-like feature surrounding IRS 16 and is produced as a result of the outflow sweeping the less dense materials in the direction away from the Galactic center and creating a dearth of emission along the western segment of the Northern Arm (Morris and Yusef-Zadeh 1987). Indeed, the timescale for sweeping out material with $n_{II} < 100\,\mathrm{cm}^{-3}$ is 50 years which is much shorter than the dynamical timescale. The radiograph presented in Figure 6 displays a larger region of Sgr A West where we note on the opposite side of IRS 16, on which Sgr A* lies, there is a chain of blobs (sources $\epsilon, \zeta, \eta$ in Yusef-Zadeh, Morris and Ekers 1990; Zhao et al. 1991) extending in an arc from Sgr A* into a "minicavity" within the Bar (see also Fig. 1). These blobs have been detected in a number of frequencies and their reality, unlike sources $\alpha$ to $\delta$ as discussed by Zhao *et al.* 1991, are well established. Figure 7 displays another rendition of the blobs at $\lambda$6cm suggestive of a gravitational perturbation of the outflow by Sgr A*.

An interesting possibility is that accretion of a portion of outflowing material from IRS 16 provides the source of luminosity for Sgr A*. VLBI measurements indicate that the elongation of Sgr A* has a position angle of $\sim 80 - 90^0$, (Lo *et al.* 1989; Jauncey *et al.* 1989) which is also consistent with recent preliminary VLBA measurements (R.L. Brown, private comm.) This may constitute morphological evidence for physical interaction with IRS 16C which is projected $1''$ away at a position angle of $90^0$ with respect to Sgr A*. The outflow should interact strongly with Sgr A* since the flow velocity $v_w$ is of order the escape velocity:

$$\frac{2GM}{rv_w^2} \approx 0.5 \left(\frac{M}{3 \times 10^6 M_\odot}\right) \left(\frac{r}{0.1\,\mathrm{pc}}\right)^{-1} \left(\frac{v_w}{700\,\mathrm{km s^{-1}}}\right)^{-2} . \tag{1}$$

where we have adopted reasonable values of $v_w$ and the distance $r$ between the two objects. The accretion need not be very efficient, since the ratio

$$\frac{L_*}{\dot{M}c^2} \approx 2 \times 10^{-4} \left(\frac{L_*}{10^7 L_\odot}\right) \left(\frac{\dot{M}}{3 \times 10^{-3} M_\odot\,\mathrm{yr^{-1}}}\right)^1 \tag{2}$$

is much less than one, where $L_*$ and $\dot{M}$ are the luminosity of Sgr A* and the mass loss rate from IRS 16 respectively. Here we have used the UV luminosity inferred from the dust emission from the circumnuclear disk (Becklin, Gatley & Werner 1982). Since this could largely be provided by early-type stars, this may be an overestimate. Ozernoy (1989) has used this fact along with the usual 10% efficiency assumed for accretion onto black holes to argue that this implies that the black hole mass is in fact only 1000 $M_\odot$. We prefer to adopt the view that the accretion is extremely inefficient. The blobs of ionized gas, assuming that they are outflowing materials moving away from Sgr A* and IRS 16, provide some support for this suggestion. Indeed, Melia (1991) has recently calculated the spectrum of synchrotron radiation emitted by spherically symmetric accretion flow onto a supermassive black hole and finds that the typical parameters for the IRS 16 outflow and Sgr A* black hole provide the luminosity and produce a reasonable spectrum. These results are suggestive but are somewhat uncertain because they rest on the assumptions about dissipation of magnetic energy as the gas and magnetic field is compressed by the flow. Future high-dynamic range VLA and VLBA maps of Sgr A* and IRS 16 will be illuminating in that they may show an interaction between these two objects.

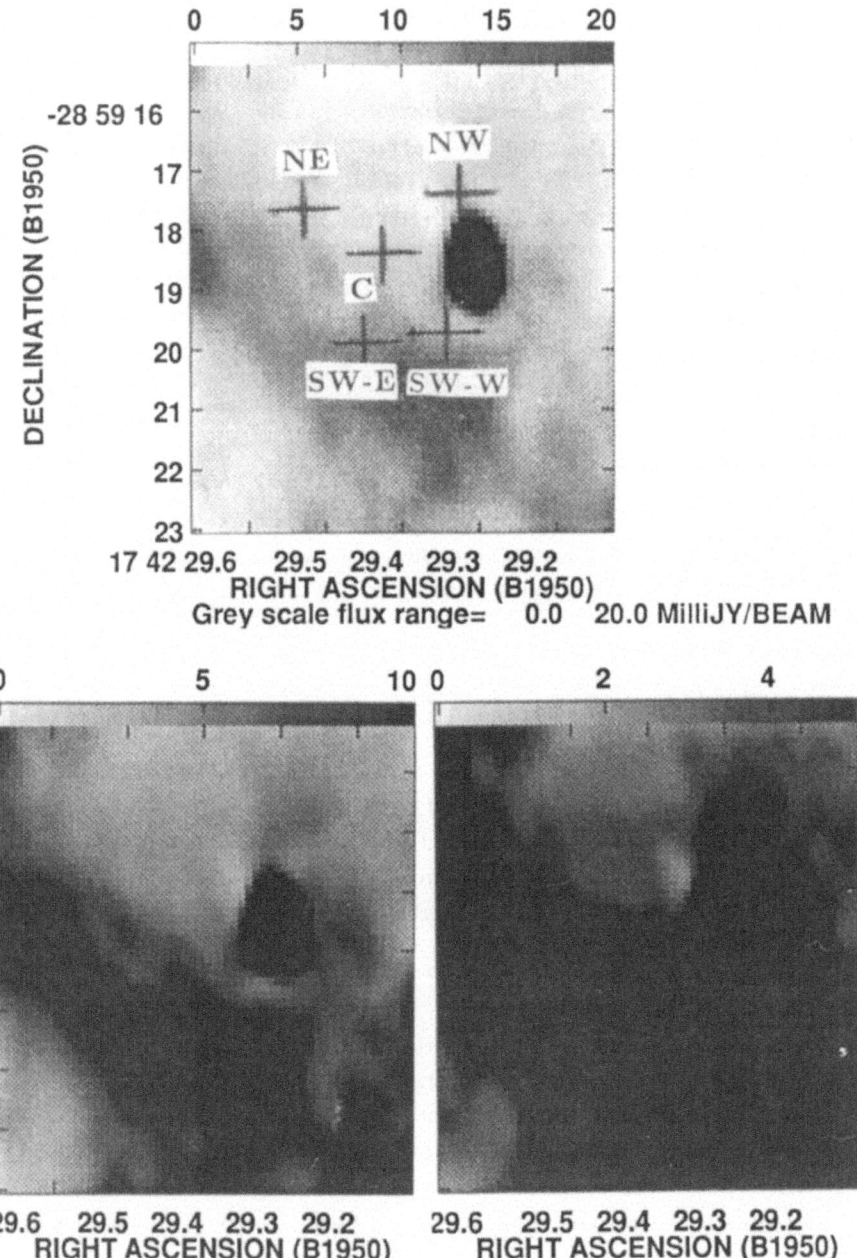

Fig. 5. Grey scale VLA images of the inner light year of the Galactic center at $\lambda 6$cm with a resolution of $0.67'' \times 0.4''$ (PA=$7^0$) with three different contrast levels. The crosses coincide with IRS 16 sources taken from Tollestrup *et al.* (1989).

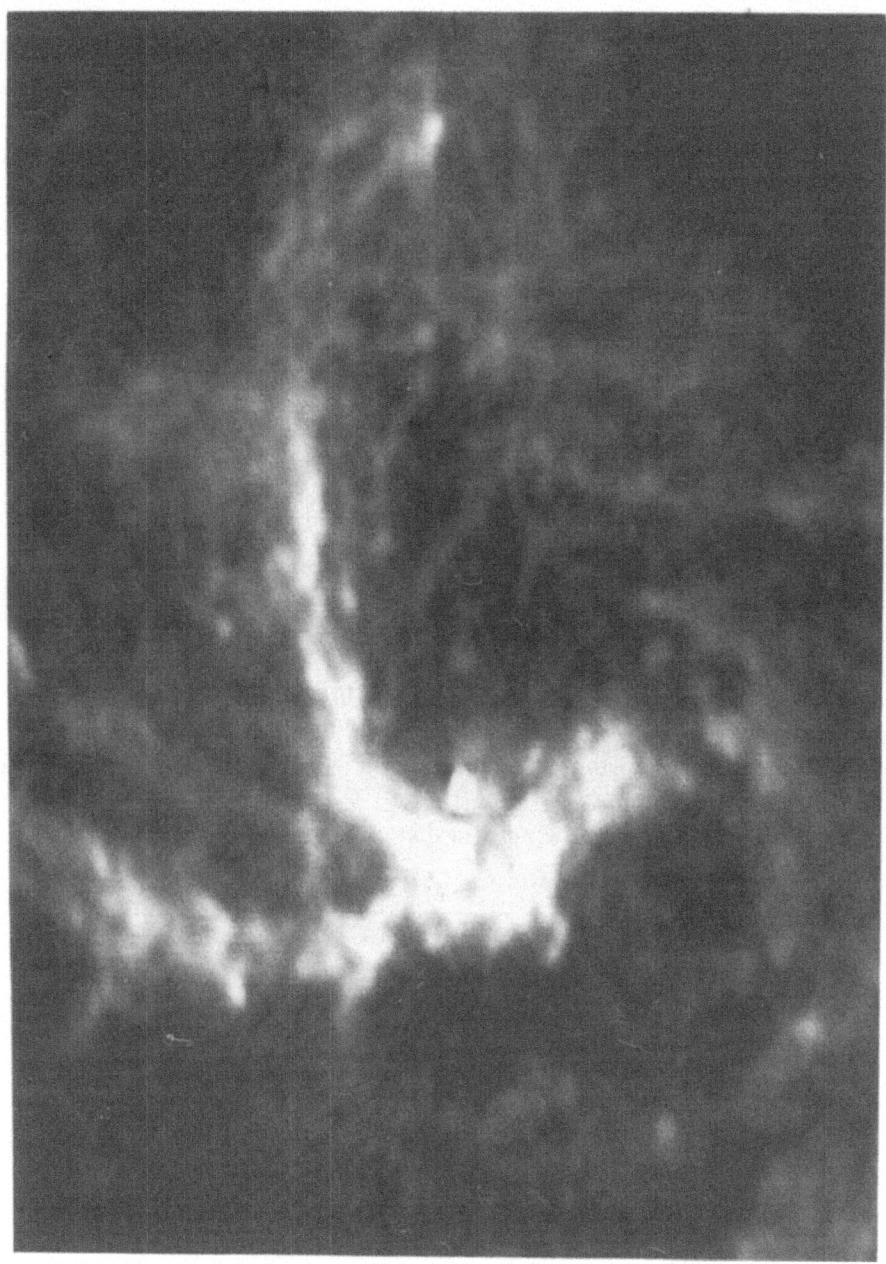

Fig. 6. This image is identical to Figure 5 except that it shows a larger region of the three Arms of Sgr A West, the Bar, the north-south dark band along the western edge of the Northern arm, IRS 7 and its tail, and Sgr A* and the adjacent blobs connecting it to the "mini-cavity" (see Fig. 1).

## 5. The Interaction with the Northern Arm

The ram pressure of the wind from IRS 16 is large enough to affect the dynamics of the gas in the arms. These interactions may prove to be a useful probe of the properties of the outflow and of the material in the arms. The velocity field, density and temperature in the arm are fairly well determined (Serabyn *et al.* 1988; Serabyn *et al.* 1991; Roberts *et al.* 1991) and we adopt $v = 200 \,\mathrm{km\,s^{-1}}$, $n_a = 10^4 \,\mathrm{cm^{-3}}$, and $T = 10^4 \,\mathrm{K}$ as typical values. The near-infrared polarization data of Aitken *et al.* (1991) indicate that magnetic field lines run along the arm. We adopt a field strength of $1 \,\mathrm{mG}$ as a reasonable estimate, consistent with the results of Killeen *et al.* (1989) for the region and previous estimates of the field strength in the Galactic center region. The Alfven speed in the arm is

$$a \approx 20 \left( \frac{B}{1\,\mathrm{mG}} \right) \left( \frac{n_a}{10^4\,\mathrm{cm^{-3}}} \right)^{-1/2} \,\mathrm{km\,s^{-1}} \tag{3}$$

so the ratio of the crossing time to the dynamical time is

$$\frac{t_a}{t_d} \approx \frac{h/a}{r/v} \approx 10\frac{h}{r}, \tag{4}$$

where $h \approx 0.1\,\mathrm{pc}$ is the width of the arm. The ratio of timescales is comparable to unity, indicating that the width of the arms is set by the balance between tidal squeezing and internal pressure. In addition, the entire width of the arm can be distorted by an interaction with the wind along the inner edge on a dynamical time.

The ram pressure of the wind at a distance $r$ from source is

$$P_w = \frac{\dot{M} v_w}{4\pi r^2} \approx 1.1 \times 10^{-7} \left( \frac{\dot{M}}{3 \times 10^{-3}\,\mathrm{M_\odot\,yr^{-1}}} \right)$$
$$\times \left( \frac{v_w}{700\,\mathrm{km\,s^{-1}}} \right) \left( \frac{r}{1\,\mathrm{pc}} \right)^{-2} \,\mathrm{erg\,cm^{-3}} \tag{5}$$

whereas the thermal and magnetic pressures of the material in the arms are

$$P_T = 2.1 n_a k T \approx 3 \times 10^{-8} \left( \frac{n_a}{10^4\,\mathrm{cm^{-3}}} \right) \left( \frac{T}{10^4\,\mathrm{K}} \right) \,\mathrm{erg\,cm^{-3}}, \tag{6}$$

and

$$P_B = \frac{B^2}{8\pi} \approx 4 \times 10^{-8} \left( \frac{B}{1\,\mathrm{mG}} \right)^2 \,\mathrm{erg\,cm^{-3}}, \tag{7}$$

respectively. The ram pressure of the wind is at least comparable to the internal pressure of the arms and can disturb them. It cannot, however completely disrupt them because the gravitational field of the central object

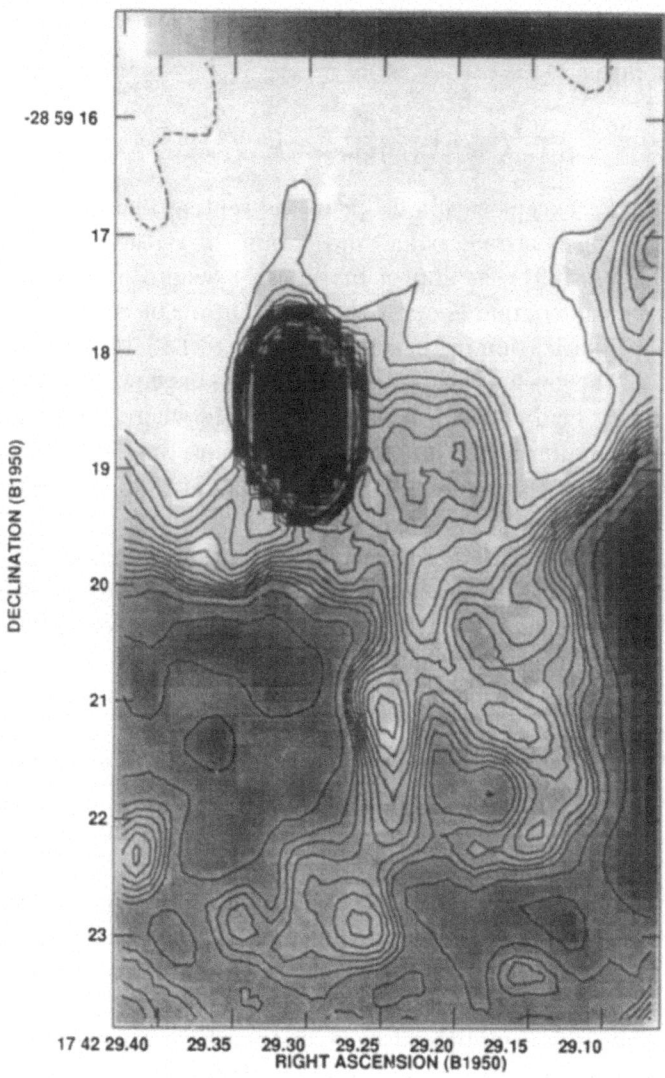

Fig. 7. This continuum VLA map at λ6cm with a resolution of $0.62'' \times 0.32''$ shows Sgr A* and its adjacent blobs which appear to be pointed toward the mini-cavity (see also Fig. 1).

is too intense. A measure of this is the ram pressure associated with the orbital motion of the gas in the arms,

$$P_g = 1.4 n_a m_H v^2$$

$$\approx 9 \times 10^{-6} \left( \frac{n_a}{10^4 \, \text{cm}^{-3}} \right) \left( \frac{v_a}{200 \, \text{km s}^{-1}} \right)^2 \text{erg cm}^{-3}, \tag{8}$$

Thus $P_g \gg P_w$ except within 0.1 pc of the center. The far-infrared polarization in the arm is strongest and most uniform at the point nearest IRS 16 (Aitken *et al.* 1991), as shown in Figure 8 where mid-IR polarization (magnetic field) is compared with the distribution of continuum emission (ionized gas). Polarization in the region closest to IRS 16 is consistent with compression of the gas and field in the arm by the outflow. One would expect that the velocity field in the arm is more chaotic where the polarization is weaker and less uniform - the magnetic field is more tangled in these regions. A comparison of existing radio and polarization data would be useful in this regard.

The waviness of the Northern Arm, as displayed in Figures 1 and 6, could arise as a result of the interaction with the outflow. Let us consider a pair of extreme situations in which the relative velocity between the outflow and the arms is directed perpendicular and parallel to the arms respectively. In the former case, it seems plausible that the ram pressure of the outflow could induce a Rayleigh-Taylor instability. Imagine a wind blowing past a linear, uniform tube of fluid, supporting it in a uniform gravitational field. It applies a uniform drag per unit length. Now buckle the tube sinusoidally in the plane containing the wind velocity. The drag now varies along the length of the tube, being largest at the points deformed in the direction of the wind and smallest at points perturbed towards the wind. The buckling will tend to increase. On the other hand, if the outflow is directed purely parallel to the arm it will tend to buckle as a result of Kelvin-Helmholtz instability driven by the shear between the two fluids. The general analysis of this instability is complicated even in the absence of rotation and the gravitational field (e.g. Hunter & Whitaker 1990), effects that are certainly important here because the observed wavelength is of order $r$. However we can consider it very roughly by using the simplest analysis for the interface between two uniform fluids in the absence of magnetic fields and thermal effects (e.g. Chandrasekhar 1961 and references therein). In that case, the timescale for growth of perturbations of wavelength $\lambda$ is

$$t_{\text{KH}} \approx \frac{2\pi u}{\lambda} \sqrt{\frac{n_a n_w}{(n_a + n_w)^2}} \tag{9}$$

where $u$ is the magnitude of the velocity difference across the interface. There is a cutoff for $\lambda/2\pi$ greater than the width of the interface. The wind density

$n_w$ is

$$n_w \approx 10 \left( \frac{\dot{M}}{3 \times 10^{-3} \, \mathrm{M_\odot \, yr^{-1}}} \right) \left( \frac{v_w}{700 \, \mathrm{km \, s^{-1}}} \right)^{-1} \left( \frac{r}{1 \, \mathrm{pc}} \right)^{-2} \mathrm{cm^{-3}}, \quad (10)$$

so that adopting $h$ as the interface width we find that

$$\frac{t_{KH}}{t_{orb}} \approx 3 \frac{\lambda}{r} \frac{v}{u} \left( \frac{r}{1 \, \mathrm{pc}} \right) \left( \frac{n_a}{10^4 \, \mathrm{cm^{-3}}} \right)^{1/2}$$

$$\times \left( \frac{\dot{M}}{3 \times 10^{-3} \, \mathrm{M_\odot \, yr^{-1}}} \right)^{-1/2} \left( \frac{v_w}{700 \, \mathrm{km \, s^{-1}}} \right)^{1/2}, \quad (11)$$

and the growth time is of order the orbital time.

This picture of the instability is very simplistic, but in any case, the gas will tend to warp as a result of the wind. It may be that the wind can provide the friction responsible for the infall of the arm towards the center. A similar idea was proposed by Quinn and Sussman (1985) who postulated that accretion was a result of drag by a static medium within the ionized cavity. In their case they required a gas density that was too high to be consistent with limits on the X-ray emission from such a medium. Consider a parcel of gas in orbit, with orbital velocity $v$. The drag force is of order $\rho_w A v \sqrt{v^2 + v_w^2}$ where $A$ is the cross sectional area of the parcel. The momentum of the parcel is $\rho v A h$, so that the ratio of the slow down time to the orbital period is

$$\frac{t_{drag}}{t_{orb}} \approx \frac{2r}{\dot{M}} \frac{\rho h v}{\sqrt{1 + (v/v_w)^2}} \approx 5 \left( \frac{10h}{r} \right) \left( \frac{n_a}{10^4 \, \mathrm{cm^{-3}}} \right) \left( \frac{v_a}{200 \, \mathrm{km \, s^{-1}}} \right)$$

$$\times \left( \frac{v_a}{200 \, \mathrm{km \, s^{-1}}} \right) \left( \frac{\dot{M}}{3 \times 10^{-3} \, \mathrm{M_\odot \, yr^{-1}}} \right)^{-1} \quad (12)$$

Although the arm is an extended feature, the frictional force per unit mass will be of the same order because of the waviness of the structure. It is possible that the outflow from IRS 16 is responsible for eating away the inner part of the circumnuclear disk in this way, creating the ionized cavity.

## 6. The Minicavity

A chain of several blobs of thermal radio continuum emission, as depicted in Figure 7, traces out an arc linking Sgr A* and the minicavity in the Bar(Yusef-Zadeh, Morris and Ekers 1990). This is a striking morphological link between the two objects and provides important clues regarding the interaction of the outflow from IRS 16 with Sgr A* and the formation of the cavity itself. This morphology strongly suggests that the cavity is a result of a blob running into the Bar. In turn, this implies that the blobs are created

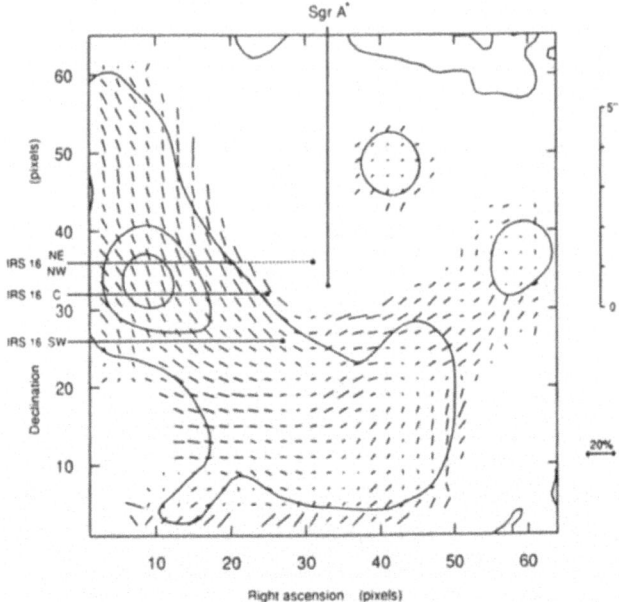

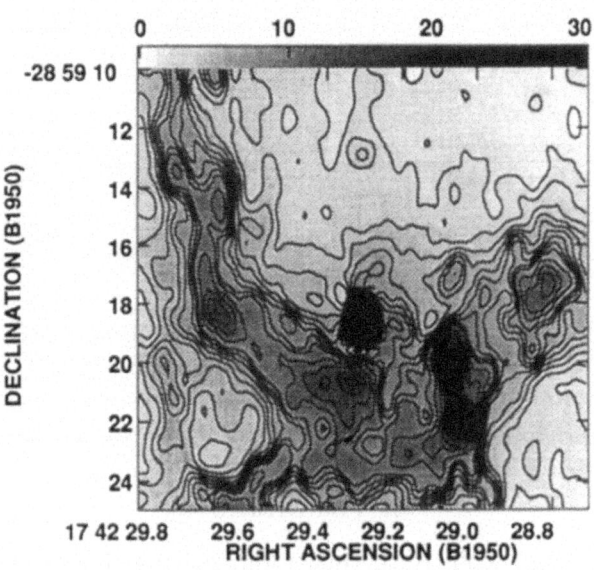

Fig. 8. 12.4 µm polarization by grain alignment is illustrated by the distribution of the magnetic field (Aitken *et al.* (1991). The bottom map shows λ6cm contours over roughly the same region.

in the vicinity of Sgr A* and are ejected, rather than representing some kind of accretion flow. Finally, because the chain lies opposite IRS 16, we suggest that the blobs are formed as a byproduct of the accretion of a fraction of the IRS 16 outflow onto Sgr A*.

The brightness temperature of the blobs is roughly 500 – 1000 K, consistent with the emission being optically thin. The emission measure then implies, assuming a temperature $T_e = 10^4$ K, a blob density of approximately $5 \times 10^4$ cm$^{-3}$, and flux density of $\approx 4.5$ mJy /beam area, as shown in Figure 7, yielding a mass of $\approx 10^{-3}$ M$_\odot$. The curvature of the arc may arise either because of the orbital motion of the source(s) of the IRS 16 wind or as a result of the gravitational attraction of Sgr A*. In either case, it implies that the blob velocity is roughly the escape velocity from Sgr A*, which is 400 km s$^{-1}$ at 0.1 pc.

The minicavity could be created by the collision of a blob with the Bar. The relative velocity between the blobs and the Bar is several hundred km s$^{-1}$, and the motion of the Bar transverse to the line of sight is probably from West to East. A collision could be expected to shock both with a postshock temperature of order $10^6$-$10^7$ K. Indeed, the western edge of the minicavity is brightest, consistent with the transverse motion of the Bar. The relative strengths of the shocks depends on the relative densities of the blob and the Bar. The emission strengths are comparable, indicating similar column densities but the blobs have smaller sizes so could well be more dense. If the density of the ambient material in the Bar were only $10^4$ cm$^{-3}$, the shock driven into the blob would be weak, with most of the dissipation ocurring in a strong, fast shock driven into the Bar. In this case, the clump lying within the minicavity could have been the one responsible for its creation.

Assuming that the timescale for a blob to traverse the length of the chain is of order the dynamical time, the mass loss rate of Sgr A* in the form of blobs is roughly one blob every hundred years which translates to $10^{-5}$ M$_\odot$ yr$^{-1}$. The rate of accretion of the wind from IRS 16 is likely to be roughly ten percent of the mass loss rate from IRS 16.3 $\times 10^{-4}$ M$_\odot$ yr$^{-1}$. The blobs may be formed as a byproduct of the accretion process, as a result of instabilities on the accretion axis, for example.

## 7. The Interaction with the Circumnuclear Disk

Figure 9 shows the intimate relationship between the ionized gas associated with Sgr A West and the surrounding molecular gas which comprises the circumnuclear disk. It has long been speculated that the molecular hydrogen emission observed from the inner part of the disk arises within shocks driven into the inner edge by an outflow from the center of the ionized cavity (Gatley et al. 1984). The ratio of the v=2-1 and 1-0 S(1) lines, which serves

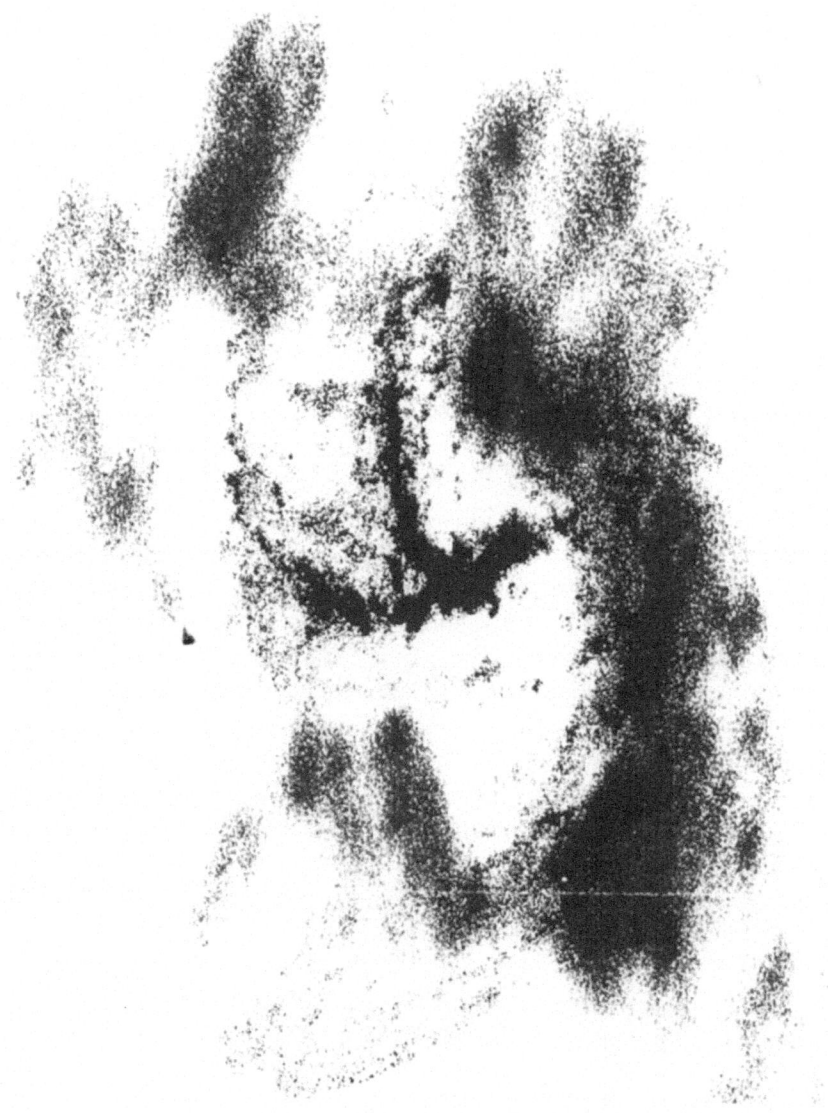

Fig. 9. HCN $1 \to 0$ emission from the circumnuclear disk is superimposed on the three arms of Sgr A West seen at $\lambda 6$ cm.

as a means to distinguish between shocked and UV heated gas is consistent with this interpretation, although it has been noted that the distinction is less clear for dense gas ($10^4 - 10^6$ cm$^{-3}$) in intense UV radiation fields (Sternberg & Dalgarno 1989). The emission shows three peaks (Figure 10), two of which are consistent with limb-brightening of the inner edges along the principal axis, and a third on the western edge of the disk. A corresponding feature on the eastern edge is noticeably absent (Gatley *et al.* 1986; DePoy, Gatley & McLean 1989).

Gatley *et al.* (1986) speculated that this resulted from an asymmetry in the outflow, caused by shielding of the eastern edge by the Northern and Eastern Arms. This idea is consistent with the scenario we have been building in this review, in which the outflow originates from several sources in IRS 16, merges and is somewhat collimated by the Bar to the South and the Northern Arm to the West (see Fig. 1). Further anisotropy is introduced by the presence of Sgr A* to the East, which gravitationally removes a significant solid angle of the outflow. The remaining avenue of escape is to the North, where the wind collides with the Western edge of the disk.

We should, however, add a word of caution. The excess of emission on the eastern edge of the cavity may reflect an excess of material there, since three similar peaks are also seen in the HCN 1-0 emission mapped by Güsten *et al.* (1988) (see the lower panel of Figure 10). The $H_2$ peak on that edge certainly implies that the outflow is present there, but the absence of emission on the opposite edge need not imply that the outflow is absent. Note, however, that the $H_2$ peak is *not* coincident with the third HCN peak and this does suggest that its position reflects the strength of the outflow rather than the amount of gas that is present.

## 8. Summary

In conclusion, we have examined the distribution and kinematics of the gas within the inner two parsecs of the Galaxy. The morphology indicates that the energetics and dynamics of gas in this region is dictated by Sgr A* and IRS 16. The unifying theme of this scenario is that IRS 16 is the source of the outflow from the central region whereas Sgr A* acts as the gravitational anchor. An appreciation of this relationship provides a framework within which we can begin to interpret the phenomenology of the region. The interaction of the wind from IRS 16 with the Bar is responsible for the high electron temperature (Roberts *et al.* 1991), the uniformity of the magnetic field, the relatively high density, the low dust-to-gas ratio (Gezari and Yusef-Zadeh 1990), and the disturbed velocity field and spatial distribution within the Bar. The interaction of the wind with Sgr A* can account for the formation of dense blobs of ionized gas, one or more of which have collided with the Bar to form the mini-cavity.

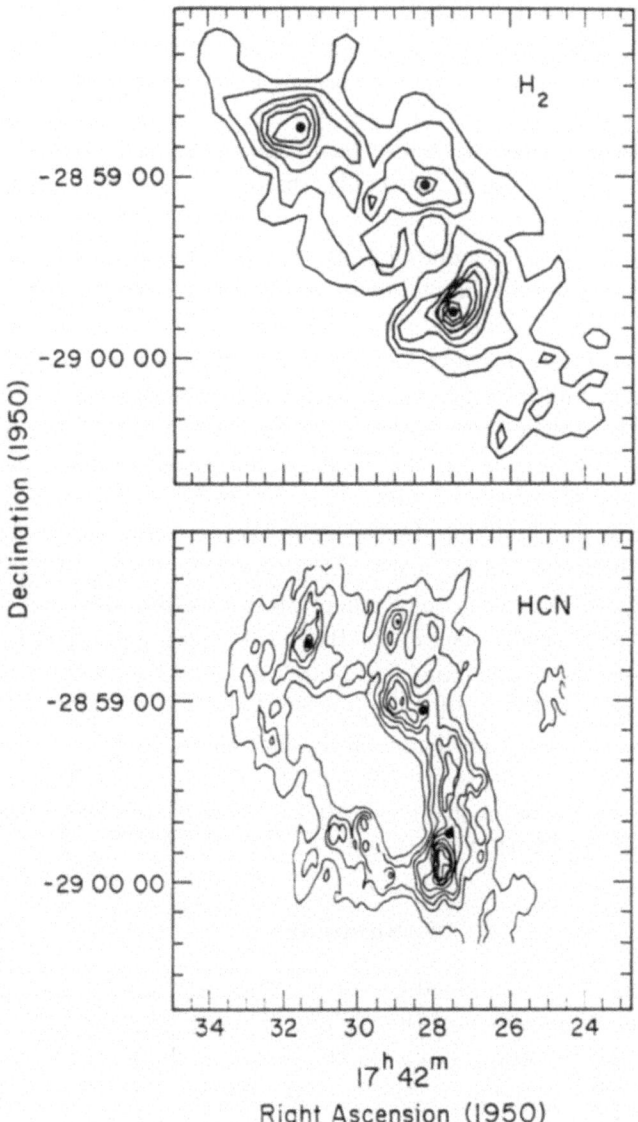

Fig. 10. Top and bottom panels show the distribution of $H_2$ v=1-0 and HCN v=1-0 emission from the circumnuclear disk (DePoy *et al.* 1989).

At larger distances from IRS 16 the wind is too weak to overcome the gravitational potential of Sgr A* but can still disturb the material in the arms. In particular, it is probably responsible for the sinusoidal structure of the Northern Arm and may provide sufficient friction to be responsible for the infall of gas from the inner edge of the circumnuclear ring towards the center. Even at these radii, the wind is powerful enough to sweep up (i) low density material to form the dark band, (ii) stellar winds to form cometary tails and (iii) to shock the inner edge of the circumnuclear disk.

# References

Aitken, D.K., Gezari, D., Smith, C.H., McCaughrean, M. and Roche, P.F. 1991, *Ap.J.*, **419**, 419.

Allen, D.A., Hyland, A.R. and Hillier, D.J. 1990, *Mon.Not.R.astr.Soc.*, **244**, 706.

Allen, D.A., and Sanders, R.H. 1986, *Nature*, **319**, 191.

Backer, D.C. and Sramek, R.A. 1982, *Ap.J.*, **260**, 512.

Becklin, E.E., Dinerstein, H., Gatley, I., Werner, M.W., and Jones, B. 1987, in *The Galactic Center: proceedings of the Symposium Honoring C.H. Townes*, ed. D.C. Backer (New York:AIP), p162.

Becklin, E.E., Gatley, I., and Werner, M.W., 1982, *Ap.J.*, **258**, 135.

Brown, R.L. and Liszt, J.H. 1984, *Ann.Rev.Astron.Astrophys.*, **22**, 223.

Chandrasekhar, S., 1961,*Hydrodynamic and Hydromagnetic Stability* (Oxford: Oxford University Press).

DePoy, D.L., Gatley, I., and McLean, I.S., 1989, *IAU Symposium No. 136, The Center of the Galaxy*, ed. M. Morris, p411.

Dressler, A., and Richstone, D.O. 1988, *Ap.J.*, **324**, 701.

Eckart, A., Hoffman, R., Duhoux, P., Genzel, R., and Drapatz, S. 1991, *The Messenger*, **65**, 1.

Ekers, R.D., van Gorkom, J.H., Schwartz, U.J. and Goss, W.M. 1983, *Astr.Astrophys.*, **122**, 143.

Gatley, I., Jones, T.J., Hyland, A.R., Beattie, D.H., Lee, T.J. 1984 *M.N.R.A.S.*, **210**, 565.

Gatley, I., Jones, T.J., Hyland, A.R., Wade, R., Geballe, T.R., and Krisciumas, K. 1986, *M.N.R.A.S.*, **222**, 562.

Geballe, T.R., Krisciunas, K., Bailey, J.A., Wade, R. 1991, *Ap.J.Lett.* **370**, L73.

Geballe, T.R., Krisciunas, K., Lee, T.J., Gatley, I., Wade, R., *et al.* 1984, *Ap.J.*, **284**, 118.

Geballe, T.R., Wade, R., Krisciunas, K., Gatley, I. and Bird, M.C. 1987, *Ap.J.* **320**, 562.

Genzel, R. 1989, *IAU Symposium No. 136, The Center of the Galaxy*, ed. M. Morris, p393.

Genzel, R. and Townes, C.H. 1987, *Ann.Rev.Astr.Ap.*, **25**, 377.

Gezari, D., and Yusef-Zadeh, F., 1990, in *Astrophysics with Infrared Arrays*, ed. R. Elston, p214.

Hall, D.N.B., Kleinmann, S.G. and Scoville, N.Z. 1982, *Ap.J.Lett.*,

Hunter Jr, J.H., and Whitaker, R.W. 1990, *Ap. J. Supp.* **71**, 777.

Jansky, K.G. 1932, *Proc. IRE*, **28**, 1920.

Jauncey, D.L. *et al.* 1991, *A.J.*, **98**, 44.

Killeen, N.E.B., Lo, K.Y., Sault, R.J., and Crutcher, R.M., 1989, *IAU Symposium 140, Galactic and Extragalactic Magnetic Fields,*, eds. R.Beck, P.P. Kronberg and R. Wielebinski, p382.

Kormendy, J. 1988, *Ap.J.*, **325**, 128.

Krabbe, A., Genzel, R., Drapatz, S. and Rotaciuc, V. 1991, *Ap.J.Lett.*, in press.

Lacy, J.H., Achtermann, J.M. and Serabyn, E. 1991, *Ap.J.Lett.*, in press.

Lacy,J.H., Townes, C.H., Geballe, T.R., and Hollenbach, D.J. 1980, *Ap.J.*, **262**, 120.

Lo, K.Y. 1989, *IAU Symposium No. 136, The Center of the Galaxy*, ed. M. Morris, p527.

McGinn, M.T., Becklin, E.E., Sellgren, K., Hall, D.N.B. 1989, *Ap.J.*, **338**, 824.

Melia, F. 1991, *Ap.J.Lett.*, submitted.

Morris, M. and Yusef-Zadeh, F. 1987, in *The Galactic Center: proceedings of the Symposium Honoring C.H. Townes*, ed. D.C. Backer (New York:AIP), p127.

Ozernoy, L. 1989, *Symposium No. 136, The Center of the Galaxy*, ed. M. Morris, p555.

Pedlar, A., Anantharamiah, K.R., Ekers, R.D., Goss, W.M., van Gorkom, J.H., Schwarz, U.J., and Zhao, J. 1989, *Ap.J.*, **342**, 769.

Quinn, P.J. and Sussman, G.J. 1985, *Ap.J.*, **288**, 377.

Rieke, G.H. and Rieke, M.J. 1989, *Ap.J.Lett.*, **344**, L5.

Roberts, D.A., Goss, W.M. and van Gorkom, J.H. 1991, *AP.J.Lett.*, **366**, L15.

Sellgren, K., McGinn, M.T., Becklin, E.E. and Hall, D.N.B. 1990, *Ap.J.*, **359**, 112.

Serabyn, E., Lacy, J.H., and Achtermann, J.M. 1991, *Ap.J.*, **378**, 557.

Serabyn, E., Lacy, J.H., Townes, C.H., and Bharat, R. 1988, *Ap.J.*, **326**, 171.

Tollestrup, E.V., Capps, R.W. and Becklin, E.E. 1989, *A.J.*, **98**, 204.

van Gorkom, J.H., Schwarz, U.J., and Bregman, J.D. 1984, in *IAU Symposium 106, the Milky Way Galaxy*, ed. H.van Woerdon, and A.J. Allen (Dordrecht:Reidel), p371.

Wardle, M. and Yusef-Zadeh, F., 1991, *Ap.J.Lett.*, in press.

Yusef-Zadeh, F., and Melia, F. 1991, *Ap.J.Lett.*, in press.

Yusef-Zadeh, F. and Morris, M. 1987, *Ap.J.*, **322**, 721. ———. 1991, *Ap.J.Lett.*, **371**, L59.

Yusef-Zadeh, F., Morris, M. and Ekers, R.D. 1989, *IAU Symposium No. 136, The Center of the Galaxy*, ed. M. Morris, p443.

———. 1990, *Nature*, **348**, 45.

Zhao, J.H., Ekers, R., Goss, M., Lo, K.Y. Narayan, R. 1989, *IAU Symposium No. 136, The Center of the Galaxy*, ed. M. Morris, p535.

Zhao, J.H., Goss, W.M., Lo, K.Y., and Ekers, R.D. 1991, *Nature*, **354**, 46.

Zhao, J.H., Roberts. D., Goss, W.M., Frail, D.A. and Lo, K.Y. 1991, *IAU Circular no. 5210*.

# MID-INFRARED EMISSION FROM THE GALACTIC CENTER

DAN GEZARI

*NASA/Goddard Space Flight Center, Code 685*
*Greenbelt, MD, 20771 USA*

**Abstract.**

New images of the central parsec of the Galaxy made with a 58 × 62 pixel array camera system at eight wavelengths between 5 and 18 $\mu$m provide a fresh perspective on physical conditions in the remarkable luminous dust and gas cloud complex at the Galactic Center. Simple array arithmetic with the images is used to calculate the color temperature and opacity of the emitting dust, and the silicate extinction distribution, with 1 arcsecond spatial resolution and sub-arcsecond astrometric accuracy. The infrared array images are compared to radio continuum maps of ionized gas emission and re-analyzed NeII velocity data. Fully sampled imaging polarization observations at 12.4 $\mu$m are used to determine the influence and spatial arrangement of individual sources near the Galactic Center. We establish a direct physical connection between dust color temperature or 12.4 $\mu$m emission features and recently detected Helium I emission line stars (Krabbe *et al.* 1992). The HeI stars are found to be co-located with the dust clouds, either as embedded luminous objects providing internal heating or as nearby external heat sources, and can account for the observed luminosity of Sgr A West. The present observational picture does not require that an exotic "central engine" be invoked to account for the infrared luminosity of the Galactic Center.

## 1. Introduction

The central parsec of the Galaxy contains a most extraordinary collection of mid-infrared (5 - 20 $\mu$m) emission sources. These objects are among the brightest of all extended Galactic mid-infrared sources. But because the Galactic Center sources are so distant (8.5 kpc) they are intrinsically much more intense and are much larger than other typical bright mid-infrared sources. For example, Galactic Center complex is 20 times more distant than the Orion BN/KL star formation region (which is comparable in angular extent and apparent brightness), thus the Galactic Center sources are intrinsically 2 - 3 orders of magnitude more intense and occupy a volume roughly 4 orders of magnitude larger.

In general, astronomical sources which are easily detected in the mid-infrared with a large ground-based telescope are few and far between. If images are made with a 2 $\mu$m array camera at random positions in the sky, several near-infrared sources (primarily visible stars) are likely to be detected in every image. However, nothing will be found in comparable 10$\mu$m images at a hundred random sky positions and probably nothing at a thousand

23

either, because detector measurements are background-limited by fluctuations in the large flux of background photons from the room-temperature telescope optics and sky. Among stellar sources, only on the order of a hundred nearby late-type stars are easily detected from the ground at 10 $\mu$m, and extended interstellar dust clouds which are sufficiently warm (T>150 $K$) and opaque ($\tau$ >0.01) to be prominent at 10 $\mu$m are not very common in the Galaxy. Thus the richness and intensity of infrared sources found at the Galactic Center suggest that exceptional physical conditions exist there.

This chapter summarizes the results of several related mid-infrared imaging projects, carried out with the array camera system. including calculation of the detailed dust color temperature, source opacity and silicate absorption distributions in the central 1 $pc$, a comparison with radio continuum emission on a 1″ scale, new imaging polarimetry measurements, a re-analysis of 12.8 $\mu$m NeII velocities, and a study of correlations between newly discovered luminous stars and features in mid-infrared emission and color temperature, which taken together provide a new view of physical conditions at the Galactic Center. Excellent reviews summarizing the extensive infrared, molecular line, and radio continuum observations of this region have been presented by Brown and Liszt (1984) and Genzel and Townes (1987). Only a brief outline of some previous work is presented here to put the new infrared imaging results in perspective.

## 2. Infrared Emission from the Galactic Center

The infrared and radio sources occupying the central few parsecs of the Galaxy make up the Sgr A West complex. Very long baseline interferometric (VLBI) continuum observations of Sgr A West at 2- and 6-cm (Lo and Claussen 1983, Yusef-Zadeh, Morris and Ekers 1989, Killeen and Lo 1989) have revealed a remarkable ionized gas cloud structure which appears very much like a system of spiral arms surrounding Sgr A*, the bright non-thermal radio point source at the nominal Galactic Center. A roughly elliptical 1′ × 2′ "molecular ring" (Gusten et al. 1987) surrounds the infrared emission region but the central parsec of the Galaxy is conspicuously devoid of molecular gas. Diffuse far infrared emission extends over several degrees along the Galactic plane near the Galactic Center (Hoffmann and Frederick 1969) peaking sharply at Sgr A West (Hoffmann, Frederick and Emery 1971). Airborne observations at 30, 50 and 100 $\mu$m with 30″ spatial resolution (Becklin, Gatley and Werner 1982) revealed a two-lobed region of far infrared emission within the central ∼2 parsecs, from a disk or torus of cool dust seen approximately edge on, corresponding in size and orientation to the ring-like molecular gas distribution. Evidence for a compact mass ∼ $10^6 M_\odot$ within the central 0.1 pc near Sgr A* was found in the velocity distribution along the arm-like structures observed in 12.8 $\mu$m Neon II

(Lacy *et al.* 1980) and in the radial dependence of a stellar mass distribution inferred from 2.3 $\mu$m CO band observations (McGinn *et al.* 1989).

There has been considerable interest in the origin of the far infrared luminosity observed at the Galactic Center. The total infrared, far infrared and submillimeter luminosity observed in the central 2 parsecs is $\sim 10^7 L_\odot$ (Low *et al.* 1969, Rieke and Low 1973, Becklin, Gatley and Werner 1982). One school of thought contends that a recent episode of massive O star formation has occured (Rieke and Lebofsky 1982, Allen *et al.* 1990) which would account for the high luminosity and ionizing radiation found in the central few parsecs of the Galaxy. Another argues that the luminosity and mass concentration is evidence for a "central engine" associated with Sgr A*, and that this object may be a massive black hole (*e.g.* Rees 1982) surrounded by an accretion disk. Radio continuum (Yusef-Zadeh, Morris and Ekers 1989) and near-infrared Brackett $\alpha$ (Forrest *et al.* 1987, Reike and Rieke 1988) observations suggest that a stellar wind from the supergiant star IRS 7 is ionized by Sgr A*. An ionized tail has been observed streaming from away from Sgr A* (Serabyn, Lacy and Achterman 1991, Yusef-Zadeh and Morris 1992), and a bow-shock at IRS 7 has also been detected by Yusef-Zadeh and Melia (1992). These observations provide additional evidence for the presence of a very energetic object at the nominal Galactic Center.

A cluster of 19 compact near- and mid-infrared "IRS" sources has been mapped in the central parsec (Becklin and Neugebauer 1975, Becklin *et al.* 1978), distributed along an extended "L" shaped ridge of mid-infrared emission called the "northern arm" and "east-west bar" by radio observers (Figures 1 - 3). Significant objects in the complex include IRS 1 and IRS 10 ($\sim$250 $K$ dense clouds and the brightest compact 10 $\mu$m peaks on the ridge), IRS 3 (a strong unresolved source similar to IRC+10216, possibly a late-type Mira star, with a deep silicate feature as is observed in the Orion BN object which absorbs $\sim$95% of its 10$\mu$m radiation), and IRS 7 (thought to be an M supergiant star of T$\sim$3500 $K$).

One of these objects in particular, IRS 16 has been widely regarded as a significant source of luminosity at the nominal Galactic Center (Becklin and Neugebauer 1975), and has been resolved into a cluster of near infrared sources within about 2″ of Sgr A* (Forrest, Pipher and Stein 1986; Tollestrup, Capps and Becklin 1989). Recent near infrared observations show that several of the IRS 16 cluster objects break up into sub-components (Allen 1989, Eckart *et al.* 1992), each of which could be a mass-losing star. However, a cluster of very luminous Helium I emission line stars has recently been discovered in the central 2 parsecs by Krabbe *et al.* (1992) which appears to be a more significant factor in the energetics of the region, as discussed in Section 7. The nature of most of the compact sources and the fundamental mechanism responsible for the high total infrared luminosity

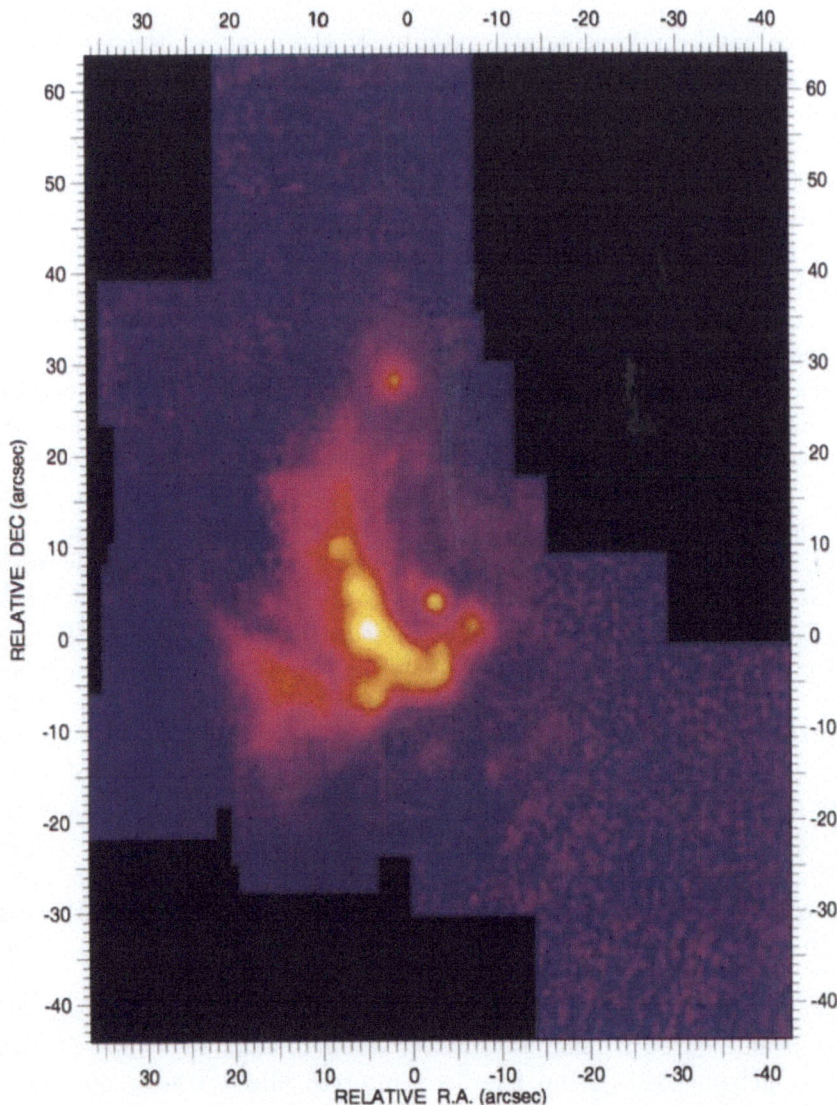

Fig. 1. 12.4 μm continuum mosaic of the Galactic Center Sgr A West complex obtained with the 58 × 62 array camera developed at NASA/Goddard Space Flight Center (Gezari *et al.* 1992) at the 3-m NASA/IRTF Telescope at Mauna Kea. The mosaic was assembled from 50 overlapping 1 min integration frames (15″ × 16″ field of view, pixel size 0.26″) which were aligned, matched and coadded to make up the final mosaic. Traces of diffuse 12.4 μm emission are seen in the area of the "western arc" radio feature (lower left), and the strongest infrared emission is very similar to the ionized gas distribution observed in 2-cm and 6-cm VLA maps of the region (see Figure 6).

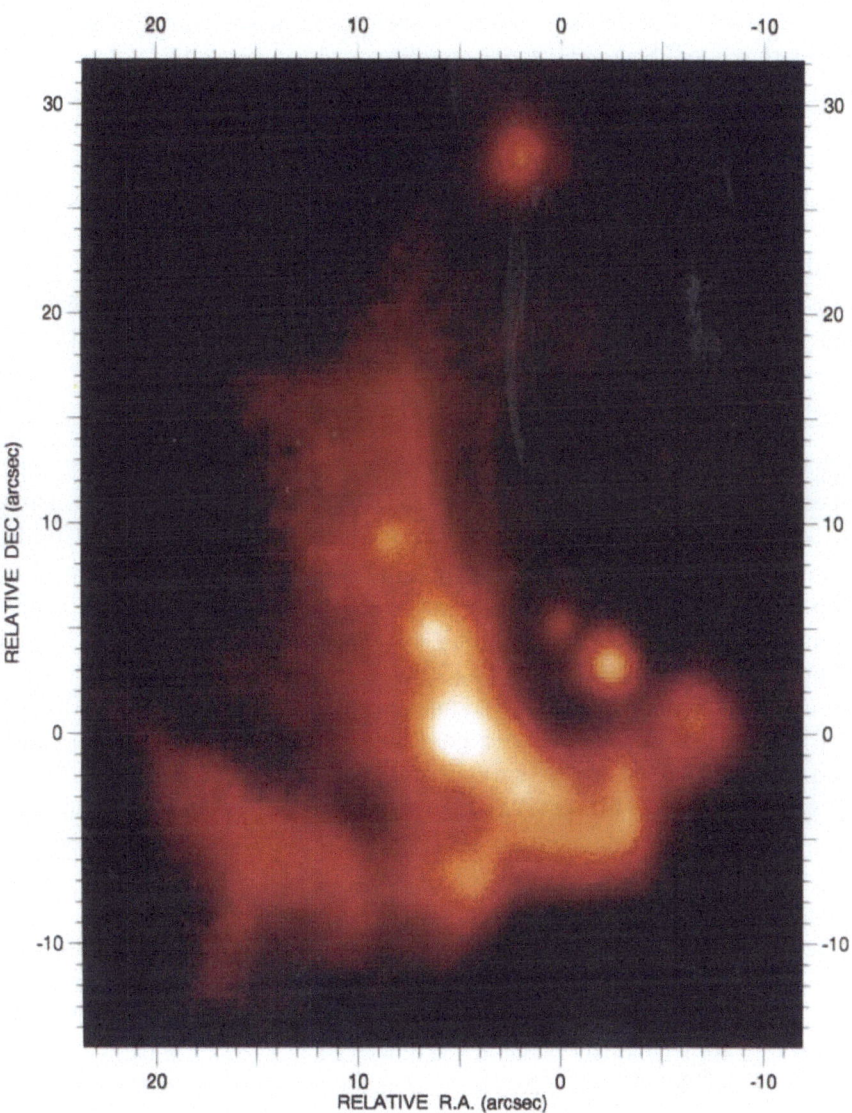

Fig. 2. The central ~2 pc of the Galactic Center Sgr A West complex, from the large mosaic image (Figure 1), imaged at 12.4 μm with our 58 x 62 array camera. The intensity display is logarithmic to show details in regions of extended faint emission. Positions are measured relative to Sgr A*.

of the Sgr A West region are still not well understood. The first mid-infrared array images of the Galactic Center were made by the author in collaboration with Rick Fienberg, Giovanni Fazio, Bill Hoffmann (Gezari et al. 1985), with a camera system developed at NASA/God- dard using an AeroJet Electrosystems 16 × 16 pixel array. Analysis of these early array observations showed that the observed 12 μm emission is due primarily to dust opacity structure, since the derived temperature peaks rather weakly at the positions of the prominent compact sources IRS 1, 5 and 10 (which seemed at that time to be among the cooler features in the region). This was interpreted as evidence that the compact sources and extended ridge are heated externally by a "central engine" near Sgr A*, rather than being heated internally by imbedded stellar luminosity sources. However, this conclusion must be now be reconsidered in light of the present higher resolution and more sensitive array camera results.

## 3. New Array Camera Results

We have made significantly improved 1.0″ resolution images of the central parsec of the Galaxy at eight wavelengths between 4.8 and 18 μm using a new 58 × 62 pixel array camera system developed in our group at Goddard (Gezari et al. 1992). The camera uses a gallium doped silicon (Si:Ga) direct readout (DRO) photoconductor detector array nominally sensitive from 5 – 17 μm, manufactured by Hughes/Santa Barbara Research Center. The array and optical assembly are cooled to 10 K. An off-axis parabolic mirror optical design produces diffraction-limited images with undetectable distortion over the 15.0″ × 16.1″ array field with good background supression. The camera incorporates a 5 - 14 μm circular variable filter ($\Delta\lambda/\lambda = 0.04$ @ 10 μm) six fixed interference filters ($\Delta\lambda/\lambda = 0.1$) between 7.8 - 12.4 μm and one at 18.1 μm. A complete description of the array camera system is presented by Gezari et al. (1992).

The Galactic Center images were made at the 3.0-meter NASA Infrared Telescope Facility (IRTF) at Mauna Kea and were nearly diffraction limited ($\lambda/D = 0.8″$ FWHM @ 10 μm). Image quality was seeing-limited at typically 1.1″ (FWHM) and relative astrometric accuracy was ±0.2 ″ . The background-limited observational system NEFD ($1\sigma$) = 0.01 Jy min$^{-1/2}$ pixel$^{-1}$ on the IRTF with a $\Delta\lambda/\lambda = 0.1$ interference filter, 0.26″ pixels, and operating at a frame rate of 30 Hz. A large-scale 12.4 μm mosaic image of the Galactic Center assembled from 50 individual array camera exposures, covering an area of 80″ × 110″ (Figure 1) shows the extended diffuse infrared structure of the Sgr A West complex. The structure of the core at 12.4 μm is shown in Figure 2, and the mid-infrared IRS sources are identified in Figure 3. An eight-color image set between 4.8 and 18.1 μm (Figure 4) has been obtained for the central 15″ array field of view (24″ = 1 pc

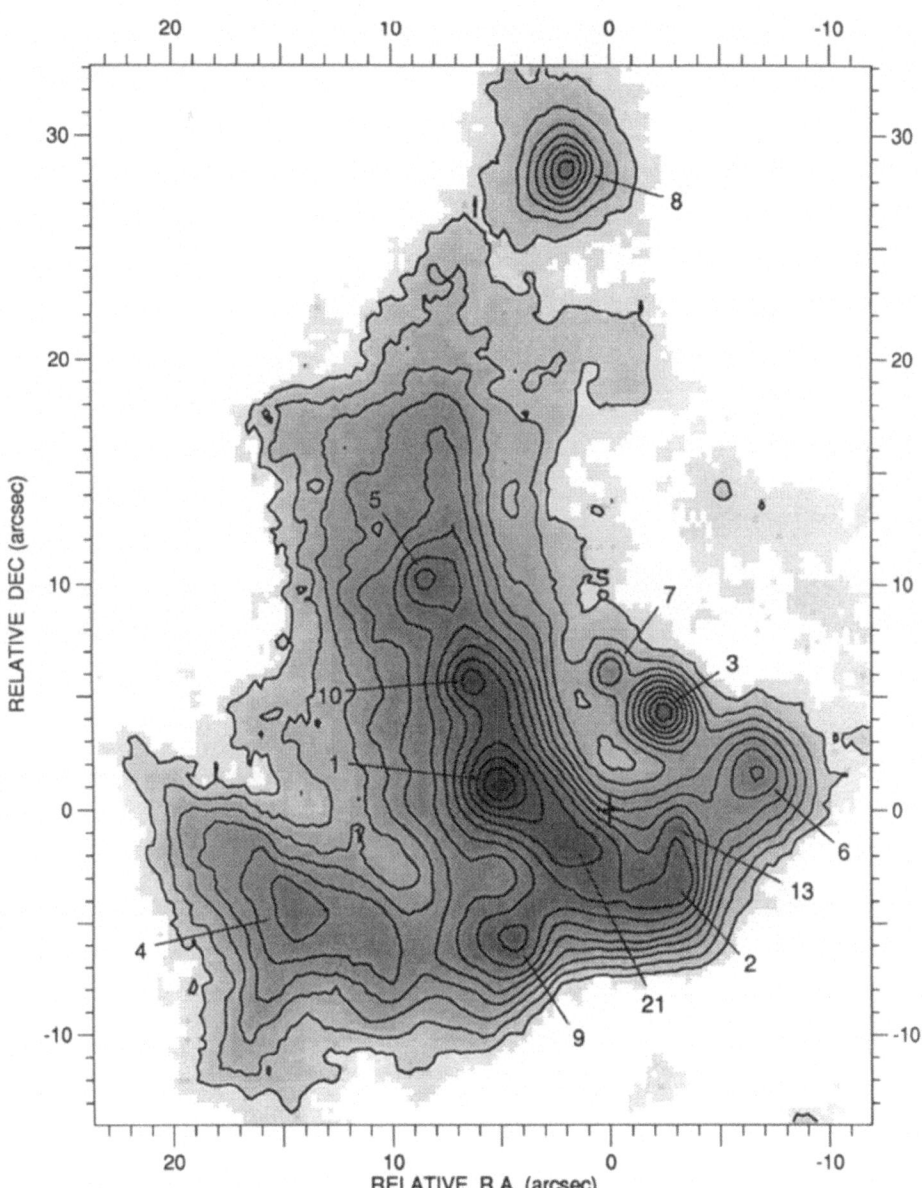

Fig. 3. Contour map of the 12.4 $\mu$m image data from Figure 2 in a linear display, plotted with 30 contour intervals between the $3\sigma$ noise level of 0.3 Jy arcsec$^{-2}$ and the peak brightness of 16 Jy arcsec$^{-2}$ at IRS 1. The prominent IRS sources are labled, and the cross shows the position of the non-thermal radio point source Sgr A$^*$ at the nominal Galactic Center. Deep integrations at 12.4 $\mu$m show no indication of infrared emission from Sgr A$^*$ ($3\sigma$ upper limit 0.1 Jy arcsec$^{-2}$ at 12.4 $\mu$m).

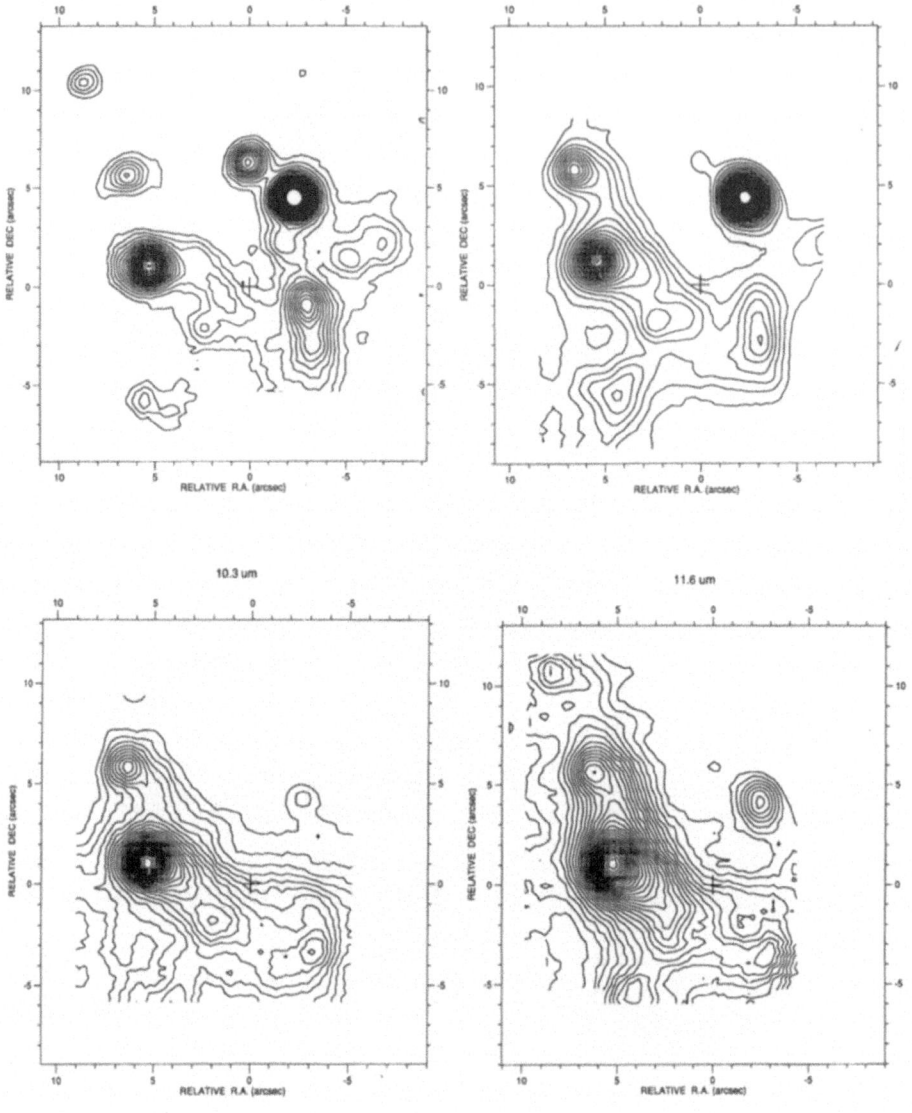

Fig. 4. This series of Galactic Center images was obtained with our 58 × 62 array camera at the 3-m NASA/IRTF Telescope at Mauna Kea (pixel size 0.26″, $\Delta\lambda/\lambda$ = 0.1). The 4.8 μm image was made at the 4-m UKIRT telescope (pixel size 0.22″, $\Delta/\lambda$ = 0.5 μm). *Top row* : *a*) 4.8 μm. The ridge-like 4.8 μm emission just east of Sgr A* *(cross)* is the remnant of the IRS 16 cluster. *b*) 7.8 μm, *c*) 8.7 μm, and *d*) 9.8 μm array images. Local extinction effects are dramatic at 9.8 μm (IRS 3 practically disappears in the display compared to the peak at IRS 1). *Bottom row* : *e*) 10.3 μm, *f*) 11.3 μm, *g*) 12.4 μm, and *h*) 18.1 μm array images. The 18.1 μm image contains contributions from the 18 μm "silicate line (in emission and absorption).

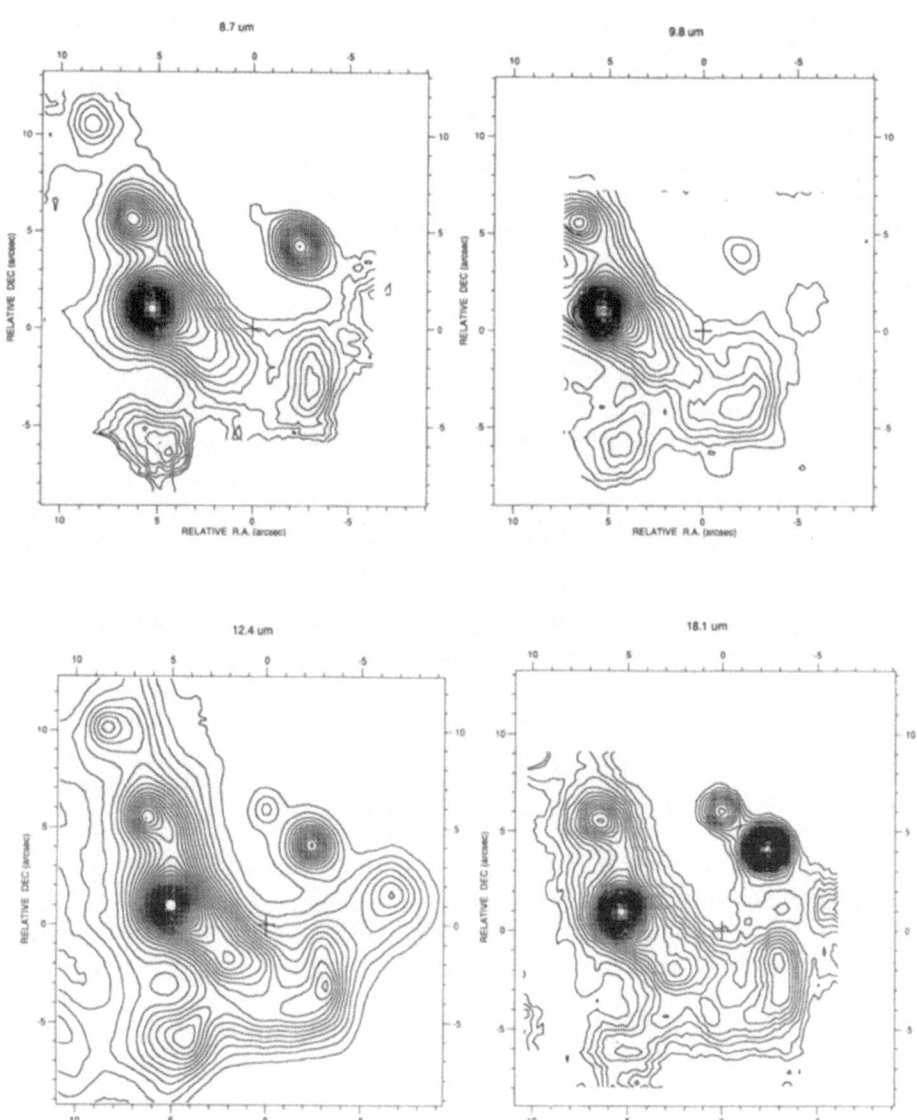

at 8.5 $kpc$), each with typically 5 minutes of integration time. Note that the images are auto-scaled to the peak brightness in each data array, thus the apparent brightness of a specific object from image to image is dependent not only on the relative intensity of sources within an image, but also on which source is strongest at each wavelength.

At 4.8 $\mu$m (Figure 4a) the hottest compact sources dominate the image and the cool, extended ridge structure which is so prominent at longer wavelengths is barely detected. The filament of 4.8 $\mu$m emission seen between IRS 1 and IRS 13 is the IRS 16 cluster. IRS 3 and IRS 13 are relatively bright, reflecting their higher color temperatures. A dramatically different source distribution is seen at 7.8 $\mu$m (Figure 4b) where cooler material in the extended ridge is becoming fully visible. IRS 3 still dominates the image. The 7.8 $\mu$m image is the longest wavelength representing uncontaminated continuum on the blue side of the deep 9.7 $\mu$m "silicate" spectral absorption feature in these sources, and can be used with some confidence as the continuum at the short wavelength end of a "stack" of multi-color image data for model analysis, as described in Section 5. At 8.7 $\mu$m (Figure 4c) the effects of silicate extinction become apparent (about a factor of 10 attenuation at 8.7 $\mu$m), dominated by the large interstellar component which is smooth across the field of view. More dramatic changes are evident at 9.8 $\mu$m and 10.3 $\mu$m (Figures 4d and 4e), and the nominal interstellar extinction correction is about a factor of 33 at 9.8 $\mu$m (Becklin et al. 1978). Extinction effects local to the sources are dramatic, particularly in the relative brightness of IRS 3 (which practically disappears in the 9.8 $\mu$m image). The detailed silicate extinction distribution derived from the images is discussed in Section 6.

At 12.4 $\mu$m (Figure 4g) the nominal interstellar extinction returns to moderate levels and the uncontaminated continuum structure of the warm dust grain distribution is fully visible. Deep array camera integrations show no indication of infrared emission from the position of the non-thermal point source Sgr A*. We have set preliminary flux density upper limits ($3\sigma$) of 0.1 $Jy$ at 12.4 $\mu$m and 0.3 $Jy$ at 18.1 $\mu$m for possible point sources at Sgr A*. These values are conservative; the Sgr A* flux densities at these wavelengths could well be an order of magnitude lower (see Zylka, Mezger and Lesch 1992 for an analysis of the spectrum of Sgr A* from 1 $\mu$m to radio wavelengths). The 18.1 $\mu$m image shows emission from the 18 $\mu$m silicate feature, but significant and intriguing differences are seen compared to the 9.8 $\mu$m image.

## 4. Spatial Analysis of the Mid-infrared Array Data

The 5 - 20 $\mu$m images can be utilized to compute "images" of derived color temperature, opacity, silicate extinction, etc. with sub-arcsecond relative astrometric accuracy and 1″ (FWHM) spatial resolution, permitting structure in the distributions to be identified with individual emission source features.

The general power law behavior of silicate, carbon or water ice grain emissivity is similar over three decades of wavelength between the near-infrared and the submillimeter. However, there are significant (order of magnitude) differences between the absorption coefficients for carbon and silicate grains, for example, in the spectral region between 5 and 20 $\mu$m (Draine and Lee 1984) to which the results are quite sensitive. A rigorous, multi-composition dust grain radiative transfer model would be required to realistically calculate the physical dust temperature, opacity in emission and extinction, and luminosity distributions over the field-of-view. A program to apply such a model to the array image data is in progress, which should soon provide an accurate quantitative picture (Gezari, Dwek and Varosi 1992) but which are not discussed here.

However, in lieu of a rigorous modeling analysis, simple array algebra performed with the image data can provide significant insight into physical conditions in the Sgr A West complex. A mathematical data "stack" can be created in which the array images are spatially aligned on common source features. The flux densities at the nine image wavelengths constitute a spectrum for any common spatial point through the stack. The color temperature distribution over the central field-of-view can be calculated from the 12.4/7.8 $\mu$m flux ratio, the relative emissive dust opacity distribution is approximated by this color temperature ratio divided by the 12.4 $\mu$m continuum emission brightness, and the distribution of cold absorbing silicate dust column density can be demonstrated by the strength of the 9.7 $\mu$m silicate spectral absorption feature (line-to continuum ratio).

The color temperature of the emitting dust sources (Figure 5a) is calculated using simple $B(\lambda)$ grey body spectra (a realistic approximation since the typical observed emissivity spectrum of interstellar grains shows that the values of the absorption coefficients are essentially the same at 7.8 and 12.4 $\mu$m). The compact sources IRS 1, 10 and 21 are seen to be local color temperature peaks of $\sim$200 $K$. Color temperatures are warmer ($\sim$230 $K$) at IRS 2 and 13, and cooler ($\sim$180 $K$) in the "northern arm" between the compact sources. Initial results of the dust modeling program show that the physical grain temperature is generally only slightly higher than the color temperature for the case of carbon grains, but roughly triple the color temperature for silicates (Gezari and Dwek 1991). The ratio = 3.5 contour ($T$=190 $K$) around IRS 1 extends west toward a temperature enhancement along the inner edge of the ridge at the position of two newly discovered luminous stars (discussed in Section 7). A slight temperature enhancement can be seen on the edge of the ridge at the position of Sgr A* which is intriguing, but not strong enough to be significant in itself. The high spatial resolution color temperature image (Figure 5a) clearly shows compact temperature peaks coinciding with all the IRS sources, contradicting our earlier conclusion (based on lower resolution images) that the IRS sources

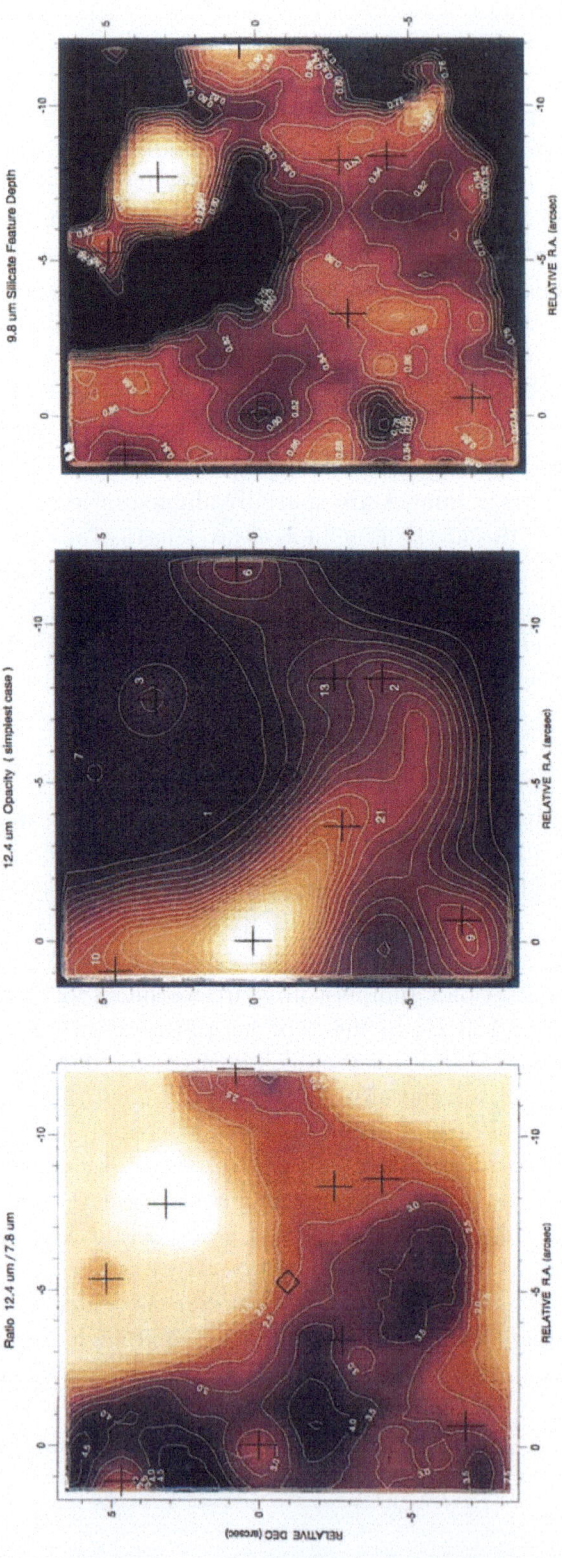

Fig. 5. *a*) Color temperature distribution in the central 0.7 *pc*, represented by the 12.4 μm /7.8 μm flux density ratio. Lighter colored regions are hotter. The positions of the IRS sources are shown (*crosses*). The ratio values correspond to color temperature, calculated from simple greybody $B(\lambda)$ spectra, where a ratio of 1.0 ~ 300 K, 2.0 ~ 230 K, 3.0 ~ 200 K, and 4.0 ~ 180 K in color temperature. A slight temperature enhancement is seen at the position of Sgr A* (*diamond*), however more dramatic correlations are seen at the positions of luminous Helium I line stars (see Figure 10). The tan colored regions of low infrared surface brightness are below the threshold of the calculation. *b*) Opacity of emitting dust grains, similar to the 12.4 μm flux density distribution, with peak opacity $\tau \sim 0.01$ occuring at IRS 1, and minimum opacity displayed of $\tau \sim 0.001$. The IRS source numbers are indicated. *c*) The 9.8 μm line/continuum ratio illustrates the distribution of silicate dust absorption. The highest silicate extinction occurs at IRS 3 where 95% of the 9.8 μm continuum flux is absorbed, and is typically near 85% on the ridge. Weakest silicate absorption is still very large, 80%, at IRS 1.

were among the cooler features in the region. The relative opacity distribution of emitting grains (Figure 5b) can be approximated in the optically thin
case by the color temperature flux ratio divided by the continuum brightness
distribution. The peak broadband opacity $\tau \sim 0.01$ at IRS 1. The opacity
distribution is much smoother than the color temperature distribution, and
quite similar to the 12.4 $\mu$m image shown for comparison in Figure 8a.
Note that IRS 2 rather than IRS 13 dominates the emission opacity in the
"east-west bar". The smoothness of the opacity distribution compared to
the much more clumpy and less dramatic temperature features reinforces
the idea that the infrared emission arises from dense dust clouds which are
heated by imbedded luminous sources.

The distribution of silicate grain absorption over the field-of-view can
be illustrated by calculating the strength of the 9.7 $\mu$m absorption feature
(Figure 5c), as the ratio of emission in the 9.8 $\mu$m filter to the unattenuated
9.8 $\mu$m continuum (obtained by interpolating between 7.8 and 12.4 $\mu$m). A
uniform component of interstellar silicate extinction is present, but spatial
variations correlated with the compact features indicate extinction effects
which are local to the sources. The effect of silicate extinction is greatest at
IRS 3, where the 9.7 $\mu$m spectral feature absorbs 95% of the interpolated
continuum, and is least at IRS 1 and 10 (but still absorbing 80% of the 9.8
$\mu$m continuum).

## 5. Comparison of Dust and Ionized Gas Emission

Our new 12.4 $\mu$m array image representing thermal emission from warm
dust (Figure 6a) can be compared with the ionized gas distribution shown
by both 2-cm VLA maps (c.f. Lo and Claussen 1983, Yusef-Zadeh, Morris
and Ekers 1989) and Br$\alpha$/Br$\gamma$ infrared images (Forrest et al. 1987). The 2-
cm radio continuum image (Figure 6b) was obtained by Yusef-Zadeh using
the VLA in its B configuration. The dust-to-gas flux density ratio (Figure 6c)
was calculated from the two data sets. The following discussion summarizes
the work of Gezari and Yusef-Zadeh (1990, 1992).

A strong similarity is seen between the general source features in the 12.4
$\mu$m mosaic image and the 2-cm VLA map. The generally good correlation
between the dust and ionized gas distributions is evidence that the warm
dust grains and ionized gas are well mixed in the central parsec of the Galaxy,
to be expected if both the radio free-free emission and infrared radiation
(from dust heated by Lyman $\alpha$ photons) are proportional to the square
of the electron number density (see also Brown and Liszt 1984). At the
same time, however, detailed comparison shows significant variation in the
12.4 $\mu$m/2-cm flux density ratio between the two extended structures, the
"northern arm" and "east-west bar" structures, generally a factor of 2 to 5
greater along the "northern arm" of Sgr A West than in the "east-west

bar" near Sgr A*. The dust-to-gas (not mass) ratio ranges from a value of $R \sim 40$ in the bar to $R \sim 150$ at IRS 1, and as high as 300 - 700 in the "northern arm" away from the compact IRS sources. There is a notably low value of the dust/gas ratio at IRS 13, which lies closest to the Sgr A* non-thermal point source. Considering the high relative positional accuracy ($\pm 0.2''$) and comparable spatial resolution ($1''$ ) of the IR array and VLA data sets, the detailed variations in the observed ratio presented here are significant.

It should be noted that there are discrepancies between the infrared and radio positions at several of the compact IRS sources, the most dramatic of which occur at IRS 1, IRS 10, and IRS 21. A similar but less obvious effect can also be seen in IRS 5. These sources all show signs of double or elongated structure in the 2-cm VLA map, and in all cases the point-like infrared sources correspond to the weaker of the two radio components. The higher value of the infrared/radio emission ratio at these sources (Figure 6c) is due to the apparent positional shift. In contrast, several other bright sources coincide in the infrared and radio, including IRS 2, IRS 4, IRS 6, IRS 9 and IRS 13. Where they occur, the displacements are generally along the ridge of the extended "northern arm" or the bar. At IRS 21 the 12.4 $\mu$m source is displaced about $1.5''$ east of the radio position. The shifts cannot be attributed to a plate scale error, which would manifest itself as radial displacements of all of the sources from IRS 7, the fiducial point. However, these displacements and the consequent lower ratio values at most of the compact IRS sources may be due to variations in the relative abundances of gas and dust, or that the infrared sources are sites of internal heating by luminous sources embedded along the ionized filaments in Sgr A West.

IRS 13 and IRS 21 lie along a $\sim 2''$ mini-cavity south of Sgr A* in the VLA radio continuum map of Sgr A West (Yusef-Zadeh, Morris and Ekers 1989). The radio components of IRS 13 and IRS 21 show a radio spectrum which is consistent with that of ionized stellar wind. Based on the morphology of ionized gas, it is plausible to consider that the relative displacement in infrared and radio peak intensities may be due to a stellar wind which has produced the mini-cavity and which has depleted the dust paricles at the inner edge of the cavity close to the source of outflow. IRS 13, on the western side of the mini-cavity, is seen to have the lowest value of the dust-to-gas emission ratio ($\sim 38$). The "east-west bar" is closer to the Galactic Center than sources in the "northern arm" and would be more likely to be directly influenced by an out-flow of material which would sweep out the dust in the region, resulting in a lower dust-to-gas ratio there compared to the "northern arm". This dust depletion could also explain the apparent relative displacement between the gas and dust peaks in the compact sources IRS 13 and IRS 21. A complete discussion of the infrared and radio results is given Gezari and Yusef-Zadeh (1992).

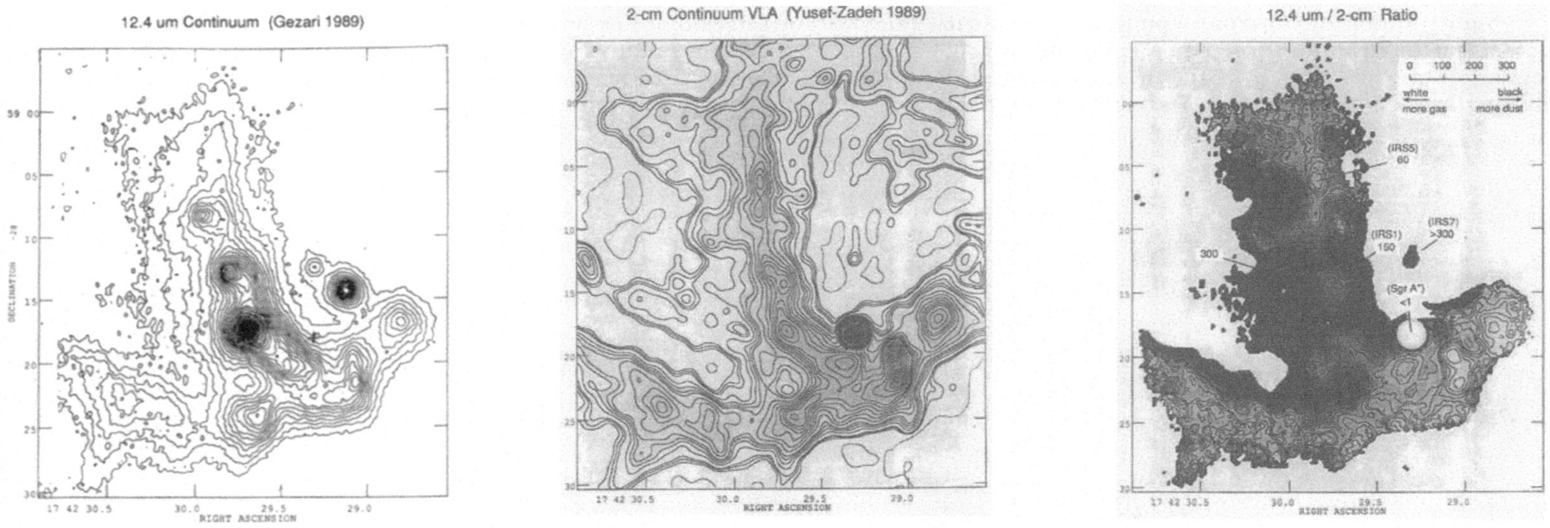

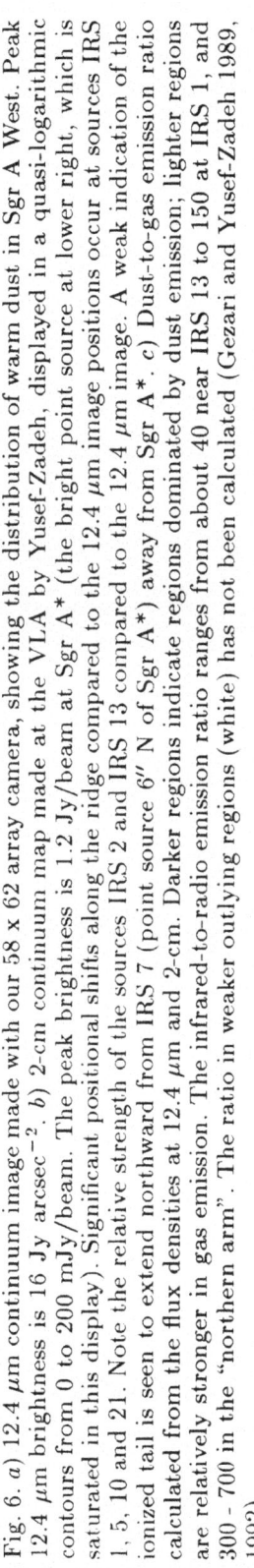

Fig. 6. *a*) 12.4 μm continuum image made with our 58 x 62 array camera, showing the distribution of warm dust in Sgr A West. Peak 12.4 μm brightness is 16 Jy arcsec$^{-2}$. *b*) 2-cm continuum map made at the VLA by Yusef-Zadeh, displayed in a quasi-logarithmic contours from 0 to 200 mJy/beam. The peak brightness is 1.2 Jy/beam at Sgr A* (the bright point source at lower right, which is saturated in this display). Significant positional shifts along the ridge compared to the 12.4 μm image positions occur at sources IRS 1, 5, 10 and 21. Note the relative strength of the sources IRS 2 and IRS 13 compared to the 12.4 μm image. A weak indication of the ionized tail is seen to extend northward from IRS 7 (point source 6″ N of Sgr A*) away from Sgr A*. *c*) Dust-to-gas emission ratio calculated from the flux densities at 12.4 μm and 2-cm. Darker regions indicate regions dominated by dust emission; lighter regions are relatively stronger in gas emission. The infrared-to-radio emission ratio ranges from about 40 near IRS 13 to 150 at IRS 1, and 300 - 700 in the "northern arm". The ratio in weaker outlying regions (white) has not been calculated (Gezari and Yusef-Zadeh 1989, 1992).

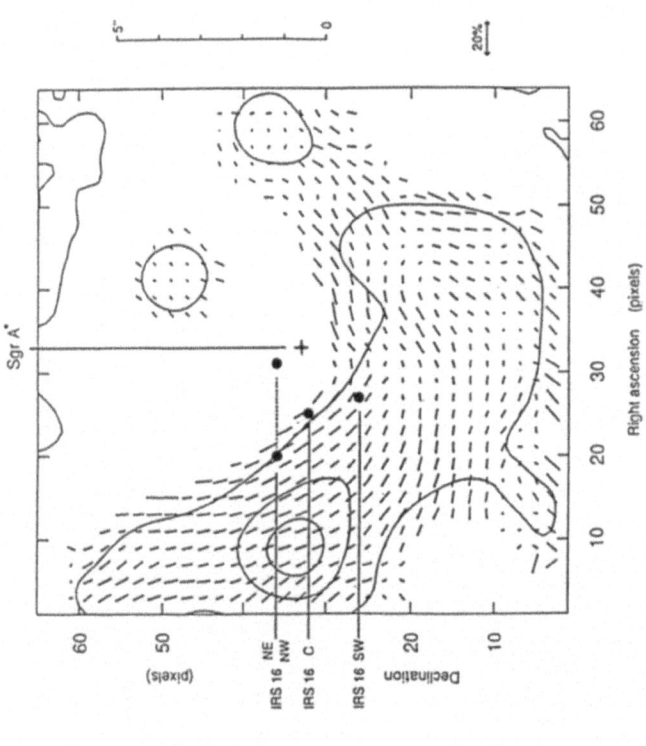

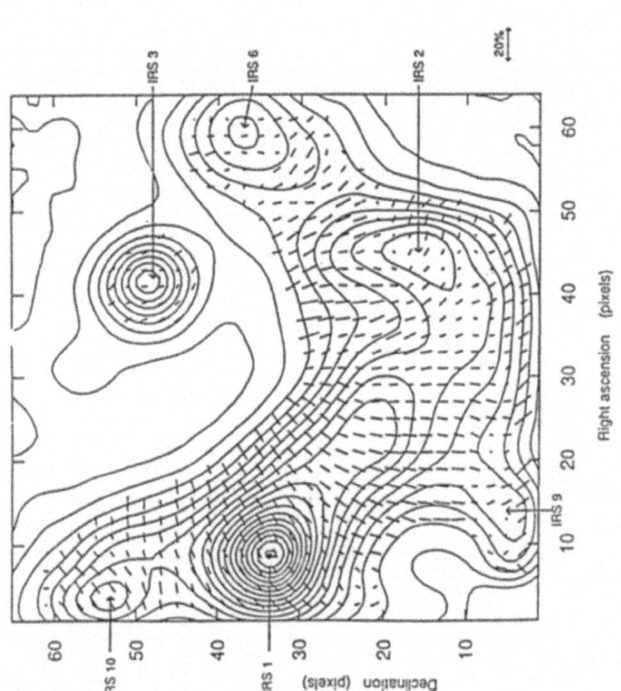

Fig. 7. *a*) 12.4 *μ*m polarization vectors for the emitting dust component (corrected for the interstellar polarization contribution), obtained at the UKIRT with our 58 × 62 array camera (Aitken *et al.* (1991). These results are free of the scattering emission component seen at near-infrared wavelengths. Maximum polarization of about 8% occurs near IRS1. *b*) Grain alignment vectors indicating the magnetic field direction (orthogonal to the polarization vectors) with maximum field strengths of roughly 10 *mG*.

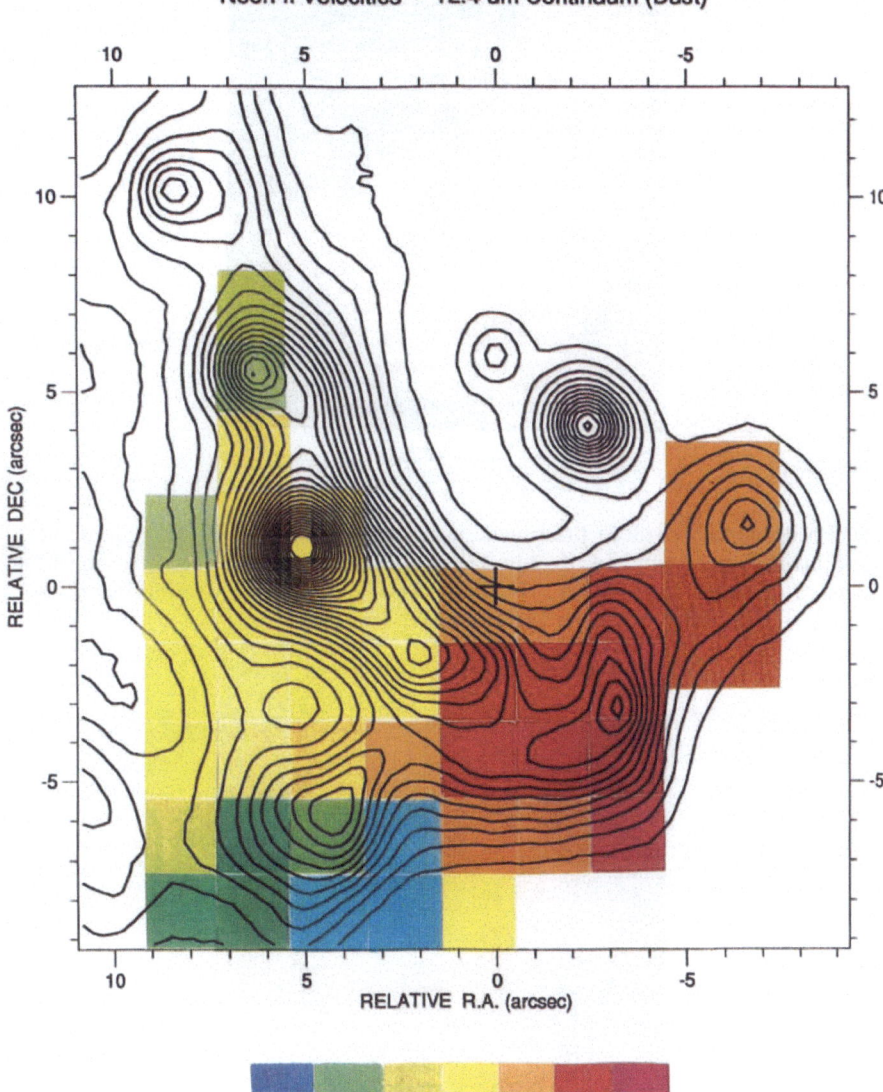

Fig. 8. Observed 12.8 $\mu$m NeII velocities ranging from +300 km s$^{-1}$ (blue) to –300 km s$^{-1}$ (red) with intensity proportional to line strength, extracted from the 2″ resolution velocity vs. declination spectra by Lacy *et al.* (1987), and overlaid on our 12.4 $\mu$m array image contours. This illustration is only schematic and not intended to be quantitatively precise; the observed velocity field is complex and not easily displayed. A 600 km s$^{-1}$ velocity gradient over 5″ is seen in the "east-west bar". The velocities over the rest of the observed region are much smaller in the "northern arm" and are close to zero (yellow) near IRS 1 and Sgr A*. The 3″ resolution data are from Lacy *et al.* (1980).

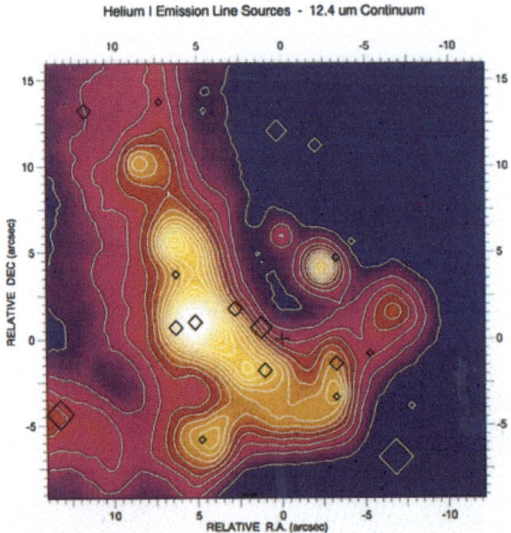

Fig. 9. Positions of the more prominent Helium I emission line stars (*diamonds*) in the central parsec found by Krabbe *et al.* (1992), shown in relation to our 12.4 μm continuum emission contours. The sizes of the symbols indicate relative source brightness. Positions are measured from Sgr A* (*cross*).

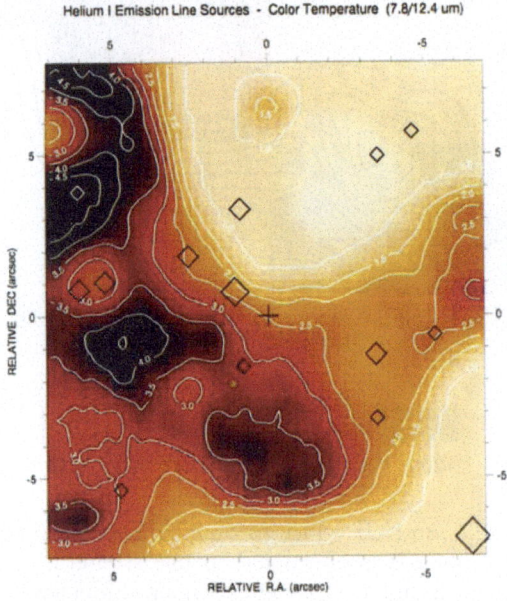

Fig. 10. Helium I star positions, plotted on the derived color temperature distrubution of warm dust in the Galactic Center (12.4/7.8 μm flux density ratio, see Figure 5a). Lighter regions are hotter. Note that the HeI stars are generally well correlated with temperature enhancements, and occasionally with temperature minima in regions depleted of dust.

## 6. Imaging Polarimetry at 12.4 $\mu$m

The first significant variations in the polarization of $\sim$10 $\mu$m radiation in the central parsec of the Galactic Center were found by Dyck, Capps and Beichman (1974), and confirmed by Capps and Knacke (1976), and Lebofsky et al. (1982). Aitken et al. (1986) contended that the mid-infrared polarization was due to alignment of warm dust grains by intrinsic magnetic fields in the Sgr A West complex, rather than by streaming of grains. A uniform line-of-sight component is present corresponding to the smooth interstellar polarization distribution observed at near-infrared wavelengths. The following discussion summarizes our 12.4 $\mu$m array polarimetry results (Aitken et al. 1991).

We have recently used the 58 × 62 array camera to make fully sampled 12.4 $\mu$m polarization images of the central 0.7 parsec at the Galactic Center. The camera was fitted with a cryogenic wire grid polarizing filter and used with a 12 $\mu$m half-wave plate mounted on the 4.0-m United Kingdom Infrared Telescope (UKIRT), controlled by the UKIRT/IRPOL polarimeter system. The results of these observations of a 14" field of view with 0.8" spatial resolution and 0.2" sampling are shown in Figure 6.

The observed polarization vectors (Figure 7a) are interpreted as the direction orthogonal to the magnetic field direction (Figure 7b). The intrinsic polarization varies slowly in amplitude and smoothly in position angle in the "northern arm" with a peak polarization of about 8% near IRS 1, decreasing to a minimum of about 2% in the "east-west bar" southeast of Sgr A*. It increases again to about 6% between Sgr A* and IRS 13 near the radio "mini-cavity" detected with the VLA by Yusef-Zadeh, Morris and Ekers (1989) with polarization angle perpendicular to the vectors near IRS 1. The regions immediately around IRS 13 and IRS 2, as well as the area north of IRS 9, seem to be nearly unpolarized. The fully sampled images show that the "east-west bar" contains distinct regions of complex polarization structure.

The observations show no evidence of disruption of the magnetic field near IRS 1 or IRS 16. Thus the results do not support the contention that there is a collision between converging filaments (Serabyn et al. 1988), causing a break and the velocity gradient observed in the ionized gas filament. The symmetry of the magnetic field vectors around the nominal Galactic Center position does not deminish the possibility that a significant mass concentration could be located near SgrA*. The lack of disruption of the polarization vectors in the "northern arm" near the position of the IRS 16 cluster suggests that IRS 16 must be separated from the arm along the line-of-sight and therefore does not lie in the plane of the arm. Thus the $10^{-3}$ $M_\odot$ yr$^{-1}$ outflow (Gatley et al. 1984) and a dominant mass concentration are unlikely to be identified with IRS 16. A complete discussion of these new imaging polarimetry results is presented by Aitken *et al.* (1991).

## 7. Neon II Line Velocities Near SgrA*

The 12.8 $\mu$m Neon II line, an important tool for high spatial resolution velocity measurements of ionized gas in visually obscured regions, was first detected in the Galactic Center by Aitken, Jones and Penman (1974; see also Aitken *et al.* 1976, Wilner 1980). The large-scale velocities observed in Sgr A West (Lacy *et al.* 1980, Serabyn and Lacy 1985) are interpreted as consistent with material primarily in the "northern arm" and "western arc" being in circular Keplerian orbit about the nominal Galactic Center. Lacy, Achtermann and Serabyn (1991) argue that this implies the presence of a $2 \times 10^6$ $M_\odot$ mass concentrated within a radius of less than 0.1 pc which is presumed to be a central black hole (Lacy, Achtermann and Serabyn 1991).

12.8 $\mu$m NeII velocity vs. declination spectra of the central 0.5 $pc$ were obtained by Lacy *et al.* (1987) using an array spectrometer with 2″ spatial resolution. To illustrate the spatial resolution of the observed velocity field, we have converted these data into a spatial map, with color depicting velocity, and intensity proportional to NeII emission line strength at those velocities. The resulting velocity "image" is superimposed on the 12.4 $\mu$m array image contours in Figure 8. It shows weighted mean velocities, and is only intended to be schematic since the velocity field is complex and not easily characterized by a simple display. It does, however, help to visualize the predominant velocity structure of the clouds. The velocity-integrated total NeII emission distribution is very similar to the 12.4 $\mu$m continuum emission distribution, so the NeII velocities can be taken as characteristic of both gas and dust cloud mass motions in the region, and the spatially resolved velocity structure can be assigned to specific infrared source features near SgrA*.

A striking and rather extreme velocity gradient is seen in the "east-west bar", with +300 km s$^{-1}$ velocity emission peaking at IRS 9, and equally strong -300 km s$^{-1}$ velocity emission near IRS 2 just 5″ away. This effect is present in the lower spatial resolution results of Lacy *et al.* 1980. The radial velocities seen over the rest of the field in Figure 8 are close to zero, particularly near IRS 1, IRS 10, IRS 21 and Sgr A*, consistent with the one-armed spiral model described by Lacy, Serabyn and Achtermann (1991). The large velocity gradient can be interpreted as rotation of the "east-west bar", moving as an isolated system (since it is not part of the one-armed spiral) about a point midway between IRS 2 and IRS 9. The rotation axis win this case lies parallel to the axis of the "northern arm," but would not pass through the nominal Galactic Center at Sgr A*. Rotation around a mass concentration at SgrA* is not ruled out, but the two high velocity clouds would be in different orbits in which case they could be moving about an arbitrarily located mass concentration. The intriguing velocity gradient in the central 0.5 pc enhances the possibility that a massive object is present near SgrA*, but unfortunately it does not further constrain the position of such an object.

The region of nearly zero 12.8 μm Ne II velocity also corresponds to the region of maximum 12 μm infrared polarization seen in our array polarimetry results, while the area of the largest positive and negative velocities shows weakest polarization, further supporting the idea that large-scale motions have disrupted the magnetic field in the "east- west" bar.

## 8. Sources of Luminosity in the Central Parsec

Krabbe *et al.* (1992) have recently detected a group of compact sources in the 2.06 μm emission line of Helium I, using an imaging array spectrometer for observations of a 40″ (2 pc) diameter field in the core of Sgr A West. They find that these objects fall into two catagories: 1) several narrow line (< 300 km s$^{-1}$) sources identified as compact mid-infrared sources, the "northern arm" and "east-west bar", and 2) a cluster of compact broad line (> 1500 km s$^{-1}$) sources identified with near-infrared continuum sources. This latter group of "HeI emission line stars", concentrated in a ∼1 pc diameter cluster centered approximately on IRS 16/Sgr A*, show an overabundance of HeI relative to HI (as indicated by the HeI/Brα, HeI/Brγ and HeI/5 GHz continuum flux ratios) compared to relative abundances in the narrow line sources, interpreted as HeI emission from the extended atmospheres of a cluster of late-type blue supergiants or Wolf-Rayet stars, with high mass-loss rates ($10^{-5}$ to $10^{-4}$ M$_\odot$yr$^{-1}$) and high outflow velocities (∼1000 km s$^{-1}$).

A detailed comparison shows significant correlations between the positions of the HeI line stars and the structure both in our 12.4 μm image (Figure 9), and in our derived dust color temperature image (Figure 10). The HeI stars coincide with the compact 4.8 or 12.4 μm sources IRS 1, 2, 3, 9, 13 and 21 in the 12.4 μm image. They are also correlated with several color temperature features not associated with the compact infrared sources. A particularly interesting spatial correlation is seen at the positions of two HeI stars next to the bright rim seen west of IRS 1 in the temperature image. These source correlations place the HeI stars physically at the Galactic Center, not simply appearing there in the line-of-sight. The fact that most of the HeI star positions coincide with compact 5 - 20 μm sources, or with color temperature enhancements, suggests that the luminous objects are embedded in the dust clouds, providing internal heating of the infrared complex. Or, in cases where the HeI line stars are located adjacent to the clouds (as they are near the edge if the ridge west of IRS1, and between IRS1 and IRS10) they appear to heat the clouds externally from dust-depleted regions.

The supergiants in this HeI star cluster have predicted individual luminosities in the range $0.1 - 2 \times 10^6$ $L_\odot$ resulting in a total luminosity of $1.2 \times 10^7$ $L_\odot$ for the entire cluster (Krabbe *et al.* 1992). Initial results of our two-component dust model fit indicate that the luminosity of all sources in the $15''$ ($\sim 0.6\,pc$) central field-of-view modeled is $1 \times 10^6$ $L_\odot$ for the carbon grain sources and $7 \times 10^6$ $L_\odot$ for silicates (Gezari and Dwek 1991), about 10% and 50% respectively of the total infrared and far-infrared luminosity observed from the Galactic Center. Thus radiation from the HeI emission line stars could account for the luminosity observed with the array camera from the central $15''$ field-of-view, and the the total luminosity observed in the far infrared from the central few parsecs of the Galaxy.

## 9. Conclusions

We have seen from these various treatments of the mid-infrared array data that: 1) the compact IRS sources are hot spots in the dense Galactic Center dust clouds, 2) the infrared/radio emission ratio shows relative dust depletion in the immediate vicinity of Sgr A* suggesting the presence of a luminous object, but the positional shift between the infrared and radio peaks is more likely to be a property intrinsic to the sources themselves, 3) the magnetic field is broken up in the "east-west bar" and IRS 16 is probably out of the plane of the extended infrared ridge structure as well as displaced from the nominal Galactic Center along the line-of-sight, 4) the high NeII velocities are not symmetrical about Sgr A*, and 5) luminous HeI emission line stars are correlated with compact infrared sources or color temperature features and must therefore be co-located with the emitting dust clouds in the Galactic Center infrared source complex.

The weight of observational evidence thus seems to be falling on the side of luminous young stars embedded within and among the dense dust clouds as the origin of the $\sim 10^7$ $L_\odot$ luminosity in Sgr A West. The present observational picture does not require that an exotic "central engine" be invoked to account for the infrared luminosity of Sgr A West. However, the nature of the extraordinary non-thermal radio point source Sgr A*, and what its relationship to a possible compact massive object at the Galactic Center might be, remain parts of a fascinating enigma.

## Acknowledgements

I am grateful to Leo Blitz for a number of stimulating discussions, to Eli Dwek, Dave Aitken and Farhad Yusef-Zadeh for highly productive collaborations, and to Walter Folz, Frank Varosi and Larry Woods for their important and on-going contributions to the infrared array camera astronomy program at Goddard. This work was supported by NASA/OSSA RTOP 188-44-23-08.

# References

Aitken, D. K., Gezari, D., Smith, C. H., McCaughrean, M., and Roche. P. 1991, *Ap. J.*, **380**, 419.

Aitken, D. K., Jones, B., and Penman, J. M. 1976, *M.N.R.A.S.*, **169**, 35p.

Aitken, D. K., Griffiths, J., Jones, B., and Penman, J. M. 1976, *M.N.R.A.S.*, **174**, 41p.

Aitken, D. K., Roche, P. F., Smith, C. H., Briggs, G. P., Hough, J. H., and Thomas, J. A. 1986, *M.N.R.A.S.*, **218**, 363.

Allen, D. A. 1989, Proc. *The Center of the Galaxy*, Ed. M. Morris, Kluwer Academic Publishers, Dordrecht, 513.

Allen, D. A., Hyland, A. R., and Hillier, D. J. 1990, *M.N.R.A.S.*, **244**, 706.

Becklin, E. E., Gatley, I., and Werner, M. W. 1982, *Ap. J.*, **258**, 135.

Becklin, E. E. and Neugebauer, G. 1968, *Ap. J.*, **194**, 265.

Becklin, E. E. and Neugebauer, G. 1975, *Ap. J.*, **200**, L71.

Becklin, E. E., Mathews, K., Neugebauer, G., and Willner, S. P. 1978, *Ap. J.*, **219**, 121.

Brown, R. and Liszt, H. 1984, *Ann. Rev. Astron. Astrophys.*, **22**, 223.

Capps, R. W., and Knacke, R. F. 1976, *Ap. J.*, **270**, 76.

Dyck, H. M., Capps, R. W., and Beichmann, C. A. 1974, *Ap. J.*, **188**, L101.

Draine. B. T. and Lee, H. M. 1984, *Ap. J.*, **285**, 89.

Eckart, A., Genzel, R., Krabbe, R., Hofmann, R., van der Werf, P. P and Drapatz, S. 1992, *Nature*, **355**, 526.

Forrest, W. J., Pipher, J. L., and Stein, W. 1986, *Ap. J. (Letters)*, **301**, L49.

Forrest, W., Shure, M., Pipher J., and Woodward, C. 1987, *The Galactic Center*, ed D. Backer, American Institute of Physics, **155**, 153.

Gatley, I., Jones, T. J., Hyland, A. R., Beattie, D. H. and Lee. T. J. 1984, *M.N.R.A.S.*, **210**, 565.

Genzel and Townes, C. H. 1987, *Ann. Rev. Astron. Astrophys.*, **25**, 377.

Gezari, D. Y. and Dwek, E. 1991, presented at IAU General Assembly XXI (Buenos Aires), Comm. 33, abstract.

Gezari, D. Y., Dwek, E., and Varosi, F. 1992, (in preparation for *Ap. J.*, ).

Gezari, D. Y., W. Folz, L. Woods and Varosi, F. 1992, *Pub. A.S.P.*, **104**, 191.

Gezari, D. Y., Tresch-Fienberg, R., Fazio, G. G., Hoffmann, W. F., Gatley. I., Lamb, G., Shu, P., and McCreight, C. 1985, *Ap. J.*, **299**, 1007.

Gezari, D. Y. and Yusef-Zadeh, F. 1990, *Astrophysics with Infrared Arrays*. A. I. P. Conference Series **13**, 214.

Gezari, D. Y. and Yusef-Zadeh, F. 1992 (in preparatiuon for *Ap. J.*, ).

Gusten, R., Genzel, R., Wright, M. C. H., Jaffe, D. T., Stutzki, J., and Harris, A. I. 1987, *Ap. J.*, **318**, 124.

Hoffmann, W. F. and Frederick, C. L. 1969, *Ap. J. (Letters)*, **155**, L9.

Hoffmann, W. F., Frederick, C. L. and Emery, R. J. 1971, *Ap. J. (Letters)*, **164**, L23.

Killeen, N. E. B. and Lo, K. Y. 1989, Proc. IAU Symposium 136 (The Center of the Galaxy), Ed. M. Morris, Kluwer Academic Publishers, Dordrecht, 453.

Krabbe, A., Genzel, R., Drapatz, S. and Rotaciuc, V. 1992, (submitted to *Ap. J. (Letters)*.

Lacy, J. H., Achtermann, J. M. and Serabyn, E. 1991, *Ap. J. (Letters)*, bf 380, L71.

Lacy, J. H., Lester, D. F., Arens, J. F., Peck, M. C., and Gaalema, S. 1987. *The Galactic Center*, D. Backer, ed. (American Institute of Physics), **155**, 142.

Lacy, J. H., Townes, C. H., Geballe, T. R., and Hollenbach, D. J. 1980, *Ap. J.*, **241**, 132.

Lebofsky, M. J., Rieke, G. H., Deshpande, M. R., and Kemp, J. C. 1982. *Ap. J.*, **236**, 672.

Lo, K. Y. and Claussen, M. J. 1983, *Nature*, **306**, 647.

Low, F. J., Kleinmann, D. E., Forbes, F. F., and Aumann, H. H. 1969. *Ap. J. (Letters)*, **157**, L97.

McGinn, M. T., Becklin, E. E., Sellgren, K., and Hall, D. N. B. 1989, *Ap. J.*, **338**, 824.

Rees, M. J. 1982, The Galactic Center, ed. G. Riegler and R. Blanford, American Institute of Physics, Conf. Series **83**, 166.

Rieke, G. H. and Lebofsky, M. J. 1982, The Galactic Center, ed. G. Riegler and R. Blanford, American Institute of Physics, Conf. Series **83**, 194.

Rieke, G. H. and Low, F. J. 1973, *Ap. J.*, **184**, 415.

Rieke, G. H. and Rieke, M. J. 1988, *Ap. J.*, **344**, L5.

Serabyn, E. and Lacy, J. H. 1985, *Ap. J.*, **293**, 445.

Serabyn, E., Lacy, J. H. and Achterman, J. M. 1991, apj **378**, 557.

Serabyn, E., Lacy, J. H., Townes, C. H., and Baharat, R. 1988, *Ap. J.*, **326**, 171.

Tollestrup, E. V., Capps, R. W., and Becklin, E. E. 1989, *A. J.*, **98**, 204.

Willner, S. P. 1976, *Ap. J.*, **206**, 728.

Willner, S. P. 1980, *Ap. J.*, **219**, 870.

Wolfire, M. G. and Churchwell, E. 1987, *Ap. J.*, **315**, 315.

Yusef-Zadeh, F. and Melia, F. 1992, *Ap. J. (Letters)*, **385**, L41.

Yusef-Zadeh, F. and Morris, M. 1992, *Ap. J. (Letters)*, in press.

Yusef-Zadeh, F., M. Morris, and R. D. Ekers 1989, *The Center of the Galaxy*, Ed. M. Morris, Kluwer Academic Publishers, Dordrecht, 443.

Zylka, R., Mezger, P. G. and Lesch, H. 1992, *Astr. Ap*, (submitted).

# THE EVOLUTION OF THE GALACTIC BULGE

R. MICHAEL RICH[1]

*Astronomy Department and Columbia Astrophysics Laboratory*
*Columbia University*
*538 West 120th Street*
*New York, NY 10027*
[1] *Alfred P. Sloan Foundation Fellow*

**Abstract.**     The central stellar bulge of the Milky Way is a distinct stellar population with age of order 10 Gyr and abundance range from -1.5 to nearly +1.0 dex. A portion of the AGB population in the bulges of the Milky Way, M31, and M32 is too luminous to have evolved from a stellar population as old as the globular clusters. Because of its distinct structure and stellar population, the history of the bulge is likely to be unique among galactic stellar populations, but may resemble that of elliptical and S0 galaxies.

## 1. Introduction

Recent images by the *COBE* satellite reveal the central bulge to be a flattened distinct component of the Milky Way. We observe a stellar population similar in appearance to the bulges of other spirals and even related to the giant ellipticals. The bulge is an experiment in galaxy evolution distinct from the well studied disk, halo, globular clusters, and Magellanic galaxies. Concepts in chemical and dynamical evolution generally applicable to ellipticals and bulges may be tested against detailed observations of the abundance and kinematics of the bulge.

Dominant views on galaxy formation and evolution still follow the rapid collapse and enrichment picture of Eggen, Lynden-Bell, and Sandage (1962). Underlying this model is the idea of an era of galaxy formation shortly after recombination, with little or no subsequent evolution. A lesson from the development of geology suggests that we benefit from some revision in our thinking.

The field of geology was revolutionized by the understanding that observable processes such as erosion and vulcanism acting over geological time could be responsible for the landforms of the present and the strata of the

47

*L. Blitz (ed.), The Center, Bulge, and Disk of the Milky Way, 47–76.*
© 1992 *Kluwer Academic Publishers.*

past. Hutton and Lyell pioneered this uniformitarian geology as an alternative to catastrophism, the viewpoint that geological evolution resulted from a small number of unusual, major events. Discovery of the expanding Universe, the cosmic microwave background, and quasars has naturally directed astronomy toward "catastrophism", emphasizing the special conditions characteristic of the early Universe.

Schweizer's (1982) assertion that the NGC 7252 merger was evolving into an elliptical introduced the ideas of uniformitarianism into galaxy evolution. In a broad sense, one must consider that agents of change observable in the present epoch–mergers, cooling flows, stripping, disk heating, and cluster relaxation–may be the *dominant* factors affecting galaxy evolution. While the importance of mergers in accounting for a large fraction of ellipticals remains an actively debated topic, it is generally accepted that the epoch of galaxy formation and evolution may extend into the present day. Color changes in a quietly evolving population alone cannot explain the presence of many blue galaxies observed at $z \approx 0.5$ (Butcher and Oemler, 1978).

The bulge of the Galaxy can be studied in sufficient detail to choose among several alternative scenarios for its formation: dissipative collapse and enrichment, a starburst terminated by an SN-driven wind, or a disk transformed into a thickened bar by secular dynamical evolution.

Given the recent revelations that the halo has no correlation between abundances and kinematics (cf. Norris, 1986; Norris and Ryan, 1989; Carney *et al.* 1990) and contains a substantial age spread (vanden Berg *et al.* 1990; Schuster and Nissen, 1989; Green and Norris, 1989) we can expect the bulge to similarly challenge our ideas about galaxy evolution.

Readers wishing an overview on the characteristics of the bulge may consult reviews by Frogel (1988) and Rich (1990a, 1991) as well as the proceedings of the 1990 ESO/CTIO meeting on *Bulges of Galaxies*. IAU Symposium 153 will also be devoted to Galactic Bulges.

We turn now in §2 to address the stellar evolution in the bulge and constraints on the bulge's age. Following, §3 discusses the abundances and chemical evolution of the bulge. The evolution of the bulge's dynamics and structure (in light of its possibly being a bar) is considered in §4. We draw from the foregoing to choose among alternative formation scenarios in §5.

## 1.1. WHAT IS THE CENTRAL BULGE ?

Until recently, we could speak of any number of populations within the central 2 kpc as properly defining the bulge. Identifcation of a metal rich population (Rich, 1988) and the 1 $\mu$m *COBE* image of the peanut-shaped bulge shifts the focus to a region within 1 kpc, where a metal rich, rotation-supported population resides.

It is true that the stellar halo extends into the inner few kpc, and it is not entirely clear how the bulge and halo population are related (as this is relevent to the reality of a gradient, it is considered in depth in §4) A globular cluster system has been identified which lies near the central kpc and is rotation supported (Armandroff, 1989). Study of the rotation field has been conducted in fields well beyond 1 kpc distant (Harding, 1990) but only interior to 1 kpc do we see the clear rotation signature (10 km/sec/deg that characterizes the true inner bulge (Catchpole, 1990; Menzies, 1990). Blanco (1988) enlists the late M giants as probes to define the extent of the bulge. The M giants virtually disappear beyond 1 kpc (or $-8°$). Figure 1 shows that the density of stars falls dramatically as one leaves the inner bulge, as Kent *et al.*'s (1991) finding of a 400 pc vertical scaleheight forcefully confirms. The M giants follow power law distributions which become increasingly steep from the M2 to M7 spectral types. The bulge also follows the $R^{-3.5}$ distribution set by the RR Lyraes (Saha, 1985).

Contained in the bulge is also the nuclear region and some $2 \times 10^8 M_\odot$ of molecular gas (Bally *et al.* 1988). It is very likely that the bulge also has hot X-ray gas (Yamauchi *et al.* 1990). As we discuss in §5, these may play a role in the ongoing evolution of the bulge.

Counts of late-type stars, K-band photometry, abundance determinations, and kinematics all confirm the distinct identity of the stellar population within 1 kpc — a "central component" (Oort, 1977; Bahcall *et al.* 1982). The origin and evolution of this population remain unaccounted for, and it is this population we shall call the bulge.

## 2. Stellar Evolution in the Bulge

The wide range of metal abundances combined with the relatively large age of the bulge stellar population produces a remarkable array of evolved progeny. In the same spatial location may be found RR Lyrae stars, late M giants, Miras, and OH/IR stars. Further, it is likely that UV-bright stars similar to those responsible for the UV rising flux in distant ellipticals are also present. Thus, the wide range in metal abundance and presence of very metal rich giants (Rich, 1988) gives rise directly to the unique stellar content of the bulge. Figure 2 qualitatively portrays this diverse population.

Evolved stellar content provides the best basis for estimating the bulge age relative to the disk and halo, shedding light on the fundamental question in its evolution.

### 2.1.1. Turnoff Photometry

Direct measurement of the age using turnoff photometry is possible beyond 1 kpc (Rich, 1985; Terndrup, 1988) but this region is at the edge of

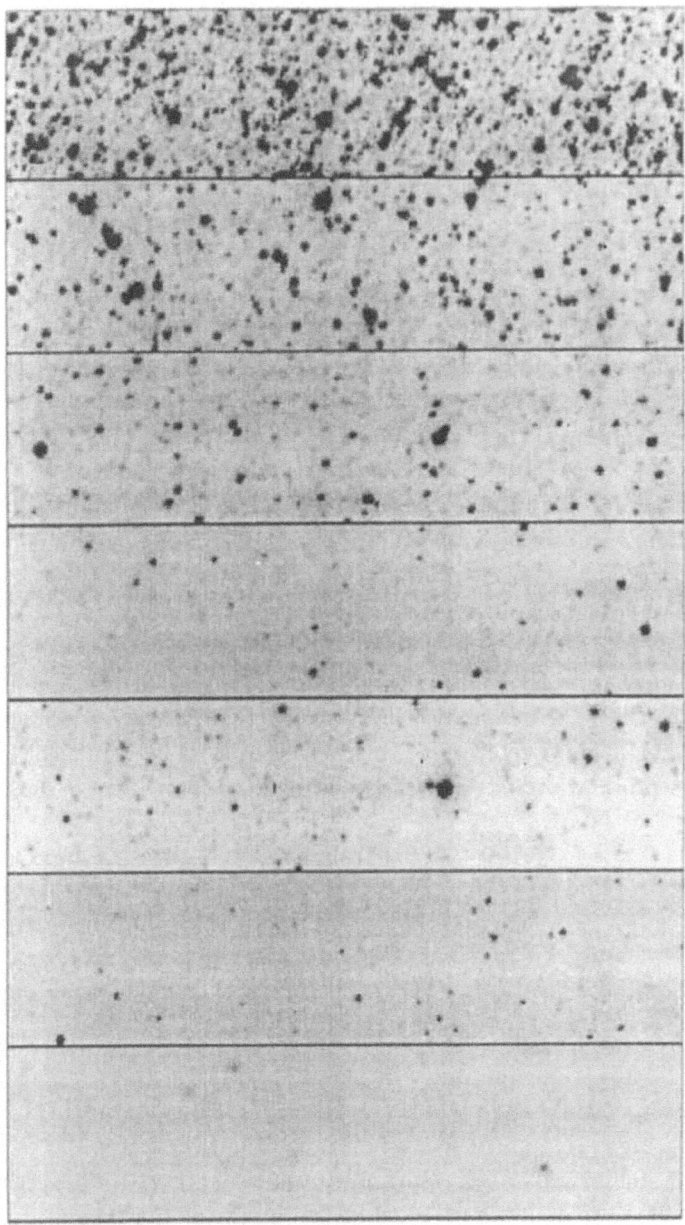

**Fig. 1.** A mosaic of bulge fields showing the dramatic decline in star density (from Tyson's 1991 Ph.D. thesis). Fields are at -2.5, -4.0, -6.0, -8.0, -10.0, -13.0, and -17.0 degrees (at 8 kpc, $1° = 140$ pc). The classical metal rich bulge population is mostly found in fields interior to 1 kpc.

# Stellar Evolution in the Bulge

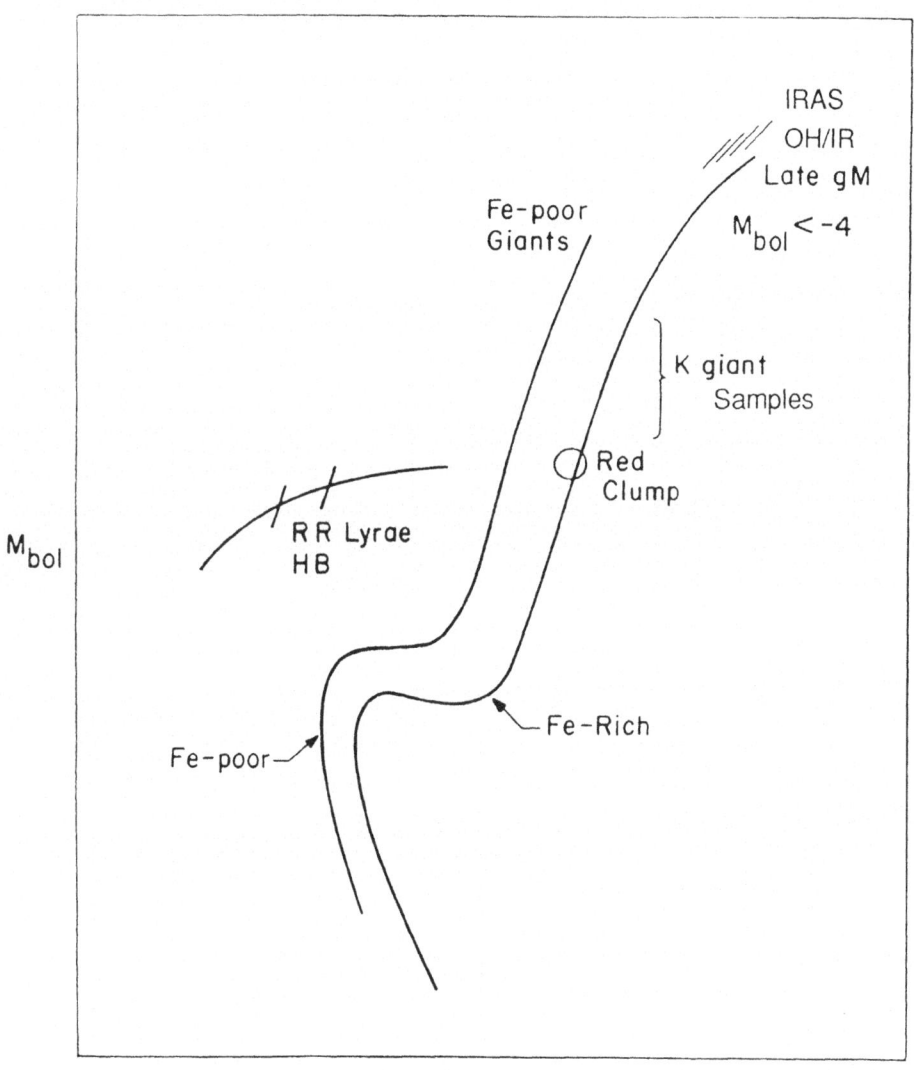

**Fig. 2.** The wide abundance range in the bulge gives rise to a varied population of evolved stars. Metal poor stars become RR Lyraes; the AGB phase of metal rich stars gives rise to Miras, IRAS, and OH/IR stars. A far-UV population (not illustrated) probably exists as well, since it is found in metal rich galaxies. Some of the most luminous AGB stars may be the progeny of turnoff stars as high as 1.25-1.5 $M_\odot$.

the classical metal rich bulge population. Revision of the bulge modulus to 8.1 kpc by Walker and Terndrup (1991) requires the isochrones in Terndrup (1988) to be faded by 0.3 mags, requiring the presence of a significant intermediate age population. However, the bulge's spatial thickness (a problem worsened by its likely bar geometry) will cause stars to be scattered brighter than the turnoff. This problem and the abundance range is inherently difficult to model. Main sequence star counts have been used to test Wood and Bessel's (1983) hypothesis that the bulge contains a substantial 1 Gyr old population. This assertion was based on pusation masses derived form long period Miras. If the Miras are indeed young, their progenitors should be found by number in the ratio $t_{MS}/t_{AGB}$. If the AGB lifetime is of order $10^5$yr, then some $10^4$ one Gyr old MS turnoff stars should be found for every Mira; Rich (1985) shows that the progenitors fall short by 3 orders of magnitude.

Because the main sequence fades only slightly for ages in excess of 10 Gyrs, one must find more precise means of distinguishing ages of old stellar populations. The evolved stellar content offers the RR Lyrae stars, which are found only in populations exceeding 10 Gyrs in age, and the luminous AGB stars which trace intermediate age populations. We turn now to examine what these stars can say about the age of the bulge.

## 2.1.2. Luminous Giants

The great surveys of Blanco (1988) delineate the bulge in its population of late M giants. Subsequently IRAS sources and OH/IR stars have defined the bulge (Habing *et al.* 1985). Such luminous giants are not generally found in the globular clusters, although luminous Miras are known in 47 Tuc and probably occur in the disk clusters.

Frogel and Whitford (1982, 1987) obtained IR photometry of the late Blanco giants and immediately learned that the giants are as luminous as $M_{bol} = -5$, a full 1.5 mags above the He core flash luminosity.

While Wood and Bessell (1983) may have overestimated turnoff masses from pulsation, it is clear that the bulge Miras have longer periods than those found in the globular clusters. A survey of IRAS Miras by Whitelock *et al.* (1991) fails to confirm > 1000 day periods initially found for the IRAS bulge stars (Harmon and Gilmore, 1987; van der Veen and Habing, 1989) but finds a peak in the period distribution at 450 days. In fact, periods as long as 770 days are discovered in the samples at $-8°$ latitude, 1000 pc from the nucleus.

If we apply a Mira P-L relation found in the LMC and bulge, we derive that the 400 day Miras have $M_{bol} < -4.5$, and the 800 day Miras lie at $-5.5$, the maximum bolometric magnitude from IR photometry of the Milky Way bulge, M31, and M32.

While it is possible that Miras evolve in luminosity and that the longest period Miras are not massive, it should be emphasized that Feast (1963) found periods and kinematics to be correlated in the sense of long period Miras being young disk objects. Feast's argument has frequently been cited in the literature; Figure 3 shows the key evidence supporting this argument which remains as compelling today as ever. Also compelling is that the bulge period distribution of Whitelock *et al.* (1991) peaks so much higher than that of the globular clusters (Figure 4). One must note with caution that there is an order of magnitude in metallicity difference between the clusters and the bulge, but the Magellanic Miras fall on the Galaxy's P-L relation despite a large metallicity difference.

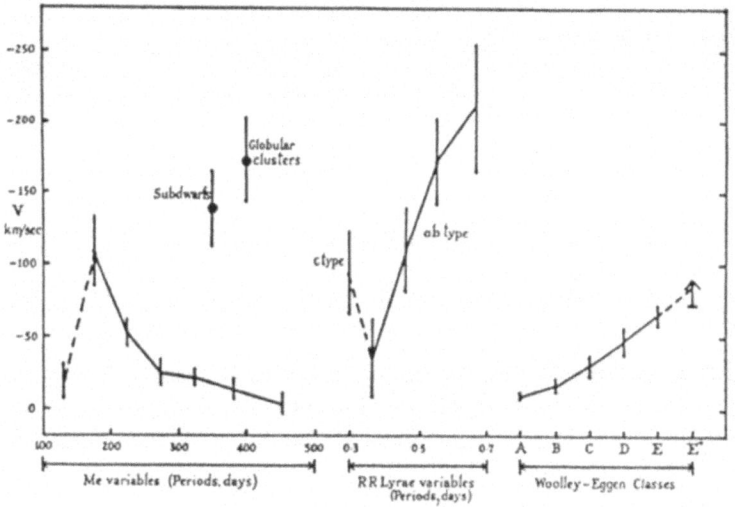

**Fig. 3.** Feast's (1963) classic figure showing how the asymmetric drift (rotation lag) depends on period for Miras and other galactic populations. Notice that Miras of period > 400 days are exclusively young disk objects.

While the P-L relationship has been questioned for long period Miras, it appears to be well established for periods of 400-500 days (cf. Whitelock *et al.* 1991) with no difference between the Magellanic Clouds and the Galaxy (no metallicity dependence). A modulus of 8kpc places the Frogel and Whitford (1987) luminosity function termination at −4.75, also in agreement with the P-L relation and the peak of the Mira period distribution.

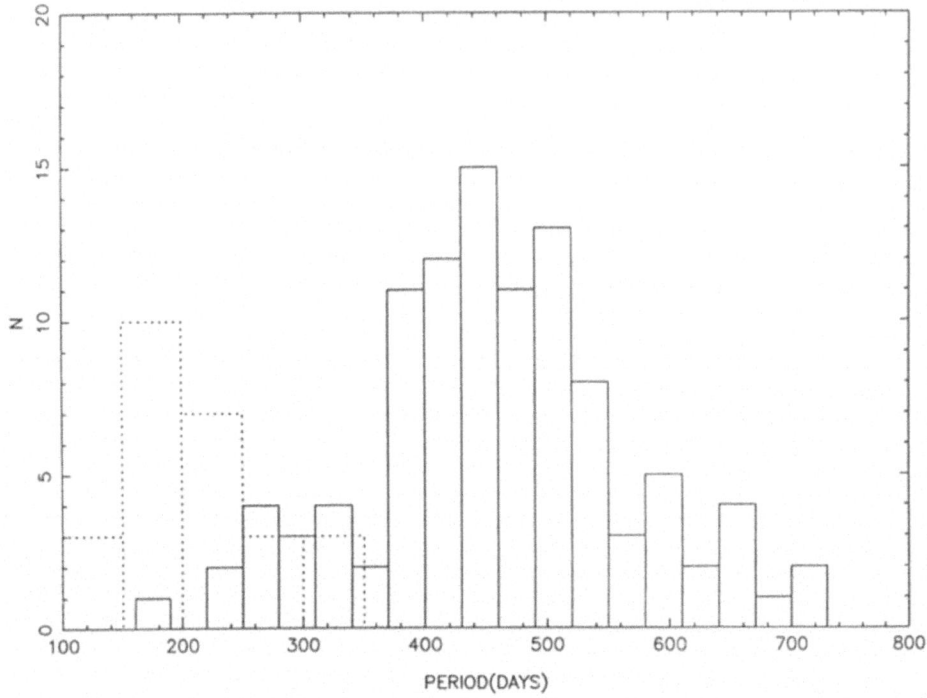

**Fig. 4.** Period distribution of globular cluster Miras (dotted line) and bulge Miras (solid line) from Whitelock *et al.* 1991. Notice that the majority of bulge Miras have longer period than the globular cluster Miras, and have periods identical to the disk Miras that Feast (1963) found to be pop I objects. The peak of the distribution is at 450 days, corresponding to $M_{bol} \approx$ $-4.8$ in the P-L relation—exactly the luminosity at which the Milky Way and M31 bulge luminosity functions terminate.

Escape from high turnoff masses might still be possible if the AGB stars are very metal rich, and $\Delta Y/\Delta Z < 1$ (cf. Renzini and Greggio, 1990). Abundance estimates for the late M giants from model atmospheres suggest a mean metal abundance of about twice solar, not large enough to propel low mass turnoff stars to the high luminosities.

Renzini and Greggio (1990) propose that the longest period Miras may be the progeny of blue stragglers. It is difficult to specify how many luminous AGB stars must be accounted for. If we assume an AGB lifetime of $6 \times 10^4 yr$ (Whitelock 1990), total bolometric bulge luminosity of $1.5 \times 10^{10} L_\odot$, the fuel consumption theorem (Renzini and Buzzoni, 1986) gives $\approx 10^4$ observed luminous AGB stars, which is the number observed. The straightforward interpretation is that the whole bulge luminosity is required but the big uncertainty is where on the luminosity function one makes the cut. Renzini and Greggio (1990) make an equally convincing argument that the extremely luminous AGB stars are blue straggler progeny. If we make the cut at higher luminosity, we must correspondingly shorten the lifetime of the evolutionary phase. The IRAS and OH/IR stars are clearly undergoing heavy mass loss. It is difficult to understand how low mass turnoff stars can attain these high luminosities and high mass loss rates. Blue straggler progeny might make the cut, but will have to successfully explain the total number of extreme stars.

An entirely different class of potentially young objects are the type II OH/IR stars (te lintel Hekkert, 1990). The population of OH/IR stars may be divided into low and high outflow velocity groups, with the cut at 27.25 km/sec. The high outflow velocity (type II) group has a significantly smaller scaleheight and greater rotation support, and is thought also (on the basis of outflow velocity) to be younger. This group is not confined to a thin disk; there are many members a few hundred pc above the plane, in the bulge.

We may ask if such luminous AGB stars are observable in the metal rich disk cluster system. If we estimate that the disk system has some $10^6$ $L_\odot$, we would find only a handful of such extreme stars. A search is warrranted because Ortolani et al. (1990) finds that the metal rich globular cluster NGC 6553 is 12 Gyr old. If such low mass stars are the progeny of 400 day Miras, we must revise our thinking about the bulge. In this author's opinion, that would be unlikely.

Finally, when one measures IR colors and makes the best estimate of bolometric luminosities, one may try to place ordinary bulge K giants on standard evolutionary tracks, such as those of Meyder and Meynet (1988). Straightforward derivation of the masses finds 2-4 $M_\odot$ for these stars. As was first pointed out by Frogel and Whitford (1982), the colors of bulge giants are too blue for their metallicities; this has subsequently been shown to be true for any color (not just IR) and remains a mystery.

### 2.1.3. The RR Lyrae Conundrum

B. Blanco's (1984) modern heroic effort to measure new periods and phases for the Baade's Window RR Lyraes set the stage for abundance measurements of these stars. The bulge RR Lyraes are certain to be more than

10 Gyr old. In fact, Lee (1991) asserts that they are the oldest RR Lyraes in the Galaxy.

Walker and Terndrup (1991) measure $\Delta S$ for about 60 of these stars and find that nearly all are about -1 dex. Lacking in the bulge is an entire population of metal rich (near solar) RR Lyraes found in the solar neighborhood. While the mechanism for producing solar abundance RR Lyraes is not known, it is potentially telling that they are too rare in the bulge (Figure 5). We may speculate that the metal rich population that produces the extreme AGB stars is too young to make RR Lyraes. It could still be as old as the old disk, but not as old as the thick disk. The value of considering the RR Lyraes is that we can discern between 10 Gyr and 13 Gyr, a difference probably too fine for bulge turnoff photometry to measure conclusively.

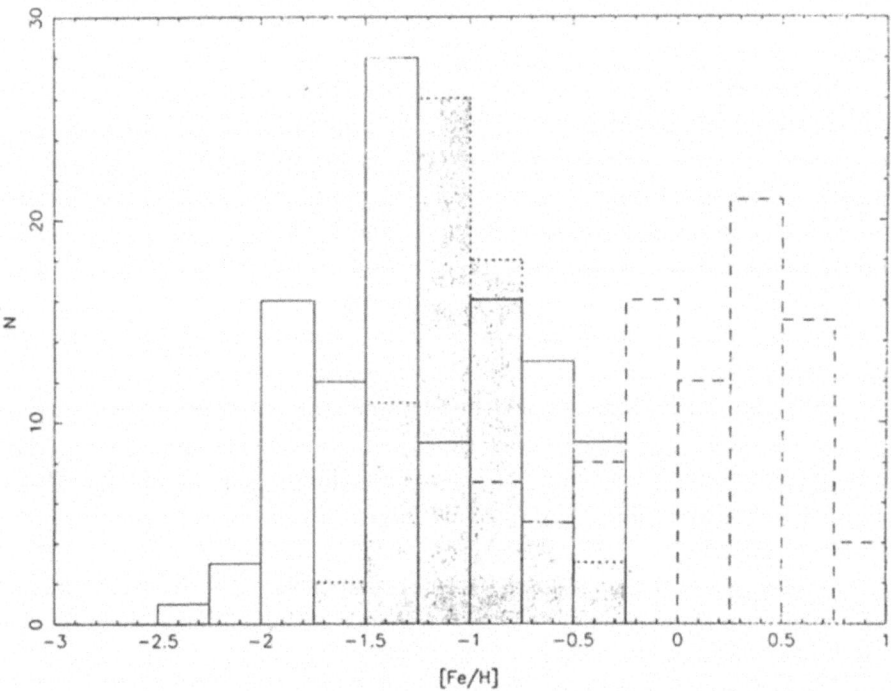

**Fig. 5.** Abundance distribution functions for *dashed line:* K giants in Baade's Window; Rich, 1988; *solid line* RR Lyraes in the solar neighborhood, from Blanco, 1992; *shaded histogram:* RR Lyraes in Baade's Window, Walker and Terndrup, 1991. Notice that the solar neighborhood RR Lyraes are more metal rich than those found in Baade's Window, despite the much higher metallicity of the galactic bulge (K giant) population. Is the metal rich bulge to young to have produced RR Lyrae stars?

Because the metal rich RR Lyraes remain unexplained, it is chancy to use them as an age constraint. However, their story is not in contradiction with inferences from the AGB.

2.2. EXTENDED GIANT BRANCHES IN M31 AND M32

Infrared arrays permit us to push the study of bulge populations to the M31 system. Figure 6 shows the clear resolution of the M31 bulge into stars (K band) within 2 arcmin ($\approx$ 400 pc from the nucleus). Rich and Mould (1991) work within 500 pc of the nucleus of M31; Freedman (1991) works similarly close to the nucleus of M32. Both groups find an extended giant branch similar to that found in the galactic bulge (see Figs. 7 and 8 below). As in the Milky Way bulge, the M31 bulge appears to extend to $M_{bol} = -5.5$, with no break at $-4.5$ as seems to occur in the Milky Way bulge fields (Frogel et al. 1990). Davies et al. (1991) employ Kent's 1989 maximum disk model to assert that the most luminous M31 stars are in fact members of an intermediate age disk population. Davies et al. (1991) conclude that the Frogel and Whitford's 1987 luminosity function applies to the M31 bulge. They successfully fit the IR photometry by using an 8 kpc modulus, which brightens the terminus to $M_{bol} = -4.75$. In fact, such a luminous termination, with an extension to $-5.0$, is quite consistent with the luminosities of the longest period (800 day) Milky Way bulge Miras derived from the P-L relation. Even if we can show that M31 and M32 are identical to the Milky Way, we must still explain the luminous AGB stars and Mira period distribution; in fact, the agreement in properties suggests that we are dealing with the same problem. Recent IR imagery of M31 by Rich, Mould, and Graham (1992) confirms that the bright AGB stars are bulges, not disk stars.

We must conclude, then, that the preponderance of evidence currently suggests that the bulge of the Milky Way and perhaps that of M31 is younger than the halo and globular clusters. It is clearly important to test this assertion by modeling the turnoff photometry and working harder to understand the AGB. Even if this is not the case, we will likely be required to modify our description of stellar evolution in metal rich stellar populations.

# 3. Chemical Evolution of the Bulge

3.1. ONE ZONE MODELS

Contrasting the disk, the bulge has a wide range of abundances extending well above the solar metallicity. The abundance distribution in Figure 9 shows the distribution is skewed to low abundances; the bulge does not have

R. M. RICH

M31 FIELD1

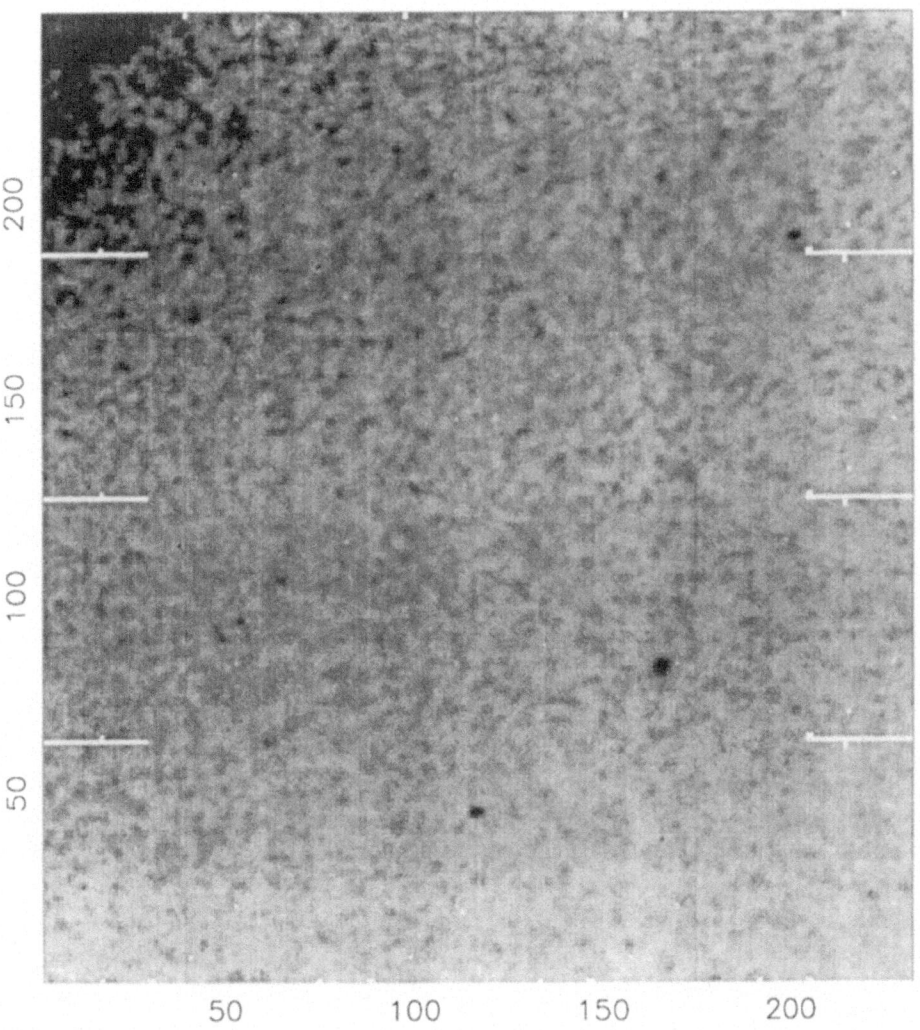

**Fig. 6.** Mosaic image of the M31 bulge, 2 arcmin SW along the major axis, in the K (2.2 micron) band (Rich, Mould, and Graham, 1992). Notice the clear resolution into individual luminous giants. The resolved stars are brighter than −4.5 bolometric magnitude, a full mag above the He core flash luminosity. Image obtained using the Caltech InSb imager at the 200-inch telescope. North at Top; East to its left.

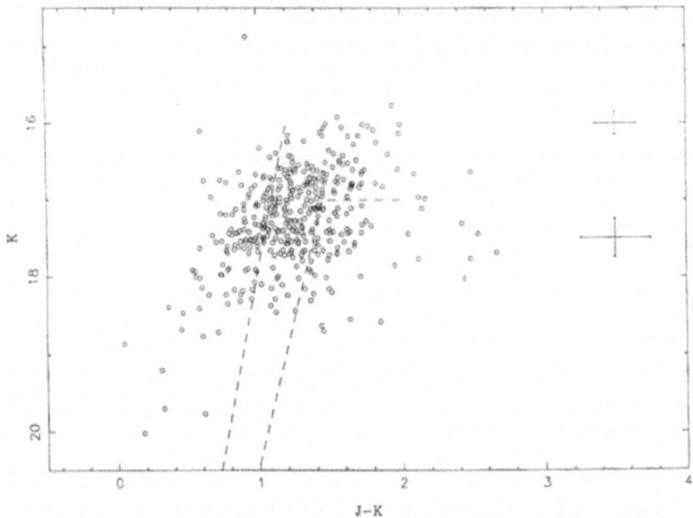

**Fig. 7.** Color-magnitude diagram of luminous giants 3 arcmin SE of the nucleus of M31, obtained with an infrared imaging array. The photometric errors are larger than 0.15 magnitude in each color. Fiducial lines indicate the extent in color and luminosity of the giant branch in the bulge of the Milky Way. Notice that the high luminosity of the giants is seen clearly. Uncertainties in bolometric corrections are larger than the photometric errors.

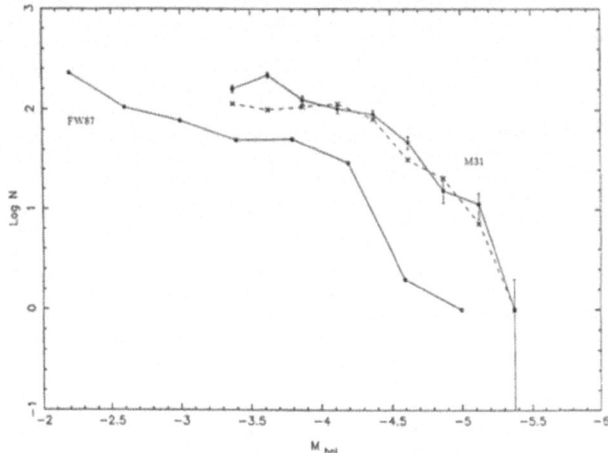

**Fig. 8.** Luminosity functions for the M31 bulge (Rich and Mould 1991), corrected for incompleteness due to crowding, compared with Frogel and Whitford's (1987) photometry in Baade's Window. Dashed connecting line is based on the color-magnitude diagram data, and the solid line is the full sample. The Milky Way luminosity function is below, lacking error flags. Milky Way stars were optically selected, and observed with a single-channel InSb photometer; M31 bulge sample is based on stars detected in the infrared K band.

a G dwarf problem. In fact, the Simple Model with a yield of twice solar fits the abundance distribution well (see Rich, 1990a).

Figures 9 and 10 show abundance distributions from Rich (1990b); these are based on low dispersion spectra and a direct comparison of bulge giant line strengths and colors with those of the globular clusters and field. The line strengths and dereddened colors given in Figure 11 show, without any calibration method, that the bulge extends to metallicities well beyond the disk clusters and the local disk field (even including metal rich giants like $\mu$ Leonis and BS 8924, and +0.4 dex). It is important to emphasize that only a handful of high dispersion abundances have been measured for bulge giants (see §3.3) and that these abundances are derived from a calibration of line strength and $(J - K)$ color in the local standards. This abundance distribution is confirmed with Washington photometry (Geisler and Friel, 1992).

Two experiments just concluding support the presence of an abundance range in the bulge, and suggest that the inner kpc has constant abundance. In collaboration with Sadler and Terndrup, I have obtained radial velocities and abundances for 500 K giants with proper motions obtained by Spaenhauer et al. 1992. The low dispersion spectra also confirm the wide abundance range in the bulge. Preliminary analysis of K giants in van den Bergh's -8° field appears to confirm the lack of a large abundance gradient. Washington photometry at several latitudes (in collaboration with N. Tyson) also fails to find strong evidence for a gradient (rather hinting at a transition). Figure 12 shows that the integrated light of SgrI and Baade's Window are indistinguishable in line strength. Terndrup et al. (1990) assert that M giant spectra may hint at a bulge/halo transition rather than a gradient. The abundances in Figure 13 from Tyson's (1991) Ph.D. thesis are quite consistent with no gradient within 1 kpc (90% of the mass).

If the bulge evolved as a simple one-zone system, then the requirements of the strict closed-box "simple" model (cf. Searle and Sargent, 1972) would be fulfilled completely. However, Zhao et al. 1991 find that if the system retains mass returned from low mass and evolved stars, this mass loss will (1) lower the abundance of the ISM at late times and (2) be available to form a generation of stars a few Gyr younger than the oldest stars in the population, but with metallicity approximately equal to the yield. Note that the last generation of stars to form is not the most metal rich. It is conceivable that the luminous giants in the bulge were formed in this way. Figure 14 shows that return from the intermediate age population would have a clearly observable effect on the metallicity distribution function, particularly if abundances can be measured to 0.1 dex accuracy.

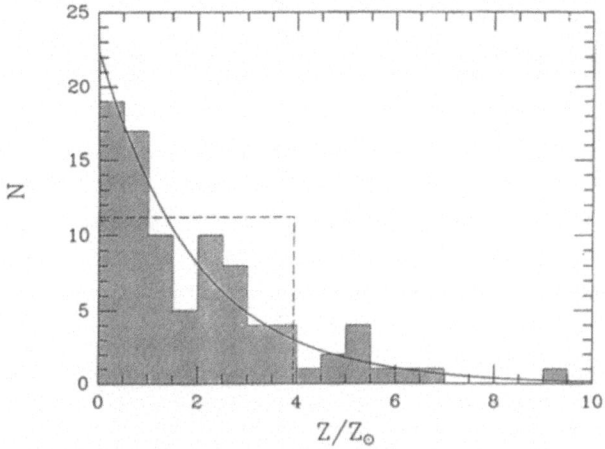

**Fig. 9.** Abundance distribution of bulge K giants measured from low dispersion spectra (Rich, 1988), fit to the Simple Model of chemical evolution, with a yield of twice solar (solid line). Dashed line indicates an unphysical model of early, complete gas exhaustion. Geisler and Friel (1992) confirm the abundance distribution with a larger sample based on the broad-band Washington photometry system. Notice that the abundance distribution is skewed toward the metal poor end.

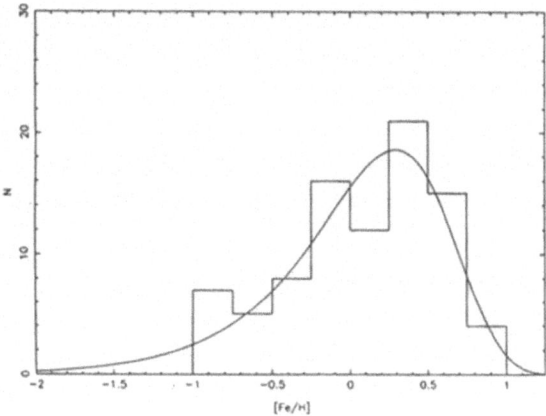

**Fig. 10.** Abundance distribution of bulge K giants as in Figure 8, in [Fe/H]. Again, the Simple Model with twice solar yield is illustrated. Notice that in logarithmic coordinates, there are very few extremely metal poor stars. If enriched gas is depleted in a steady wind, the peak of the abundance distribution is shifted to lower yield, but the shape of the distribution does not change. It is interesting to note that the halo can be considered to have a very low yield, requiring no pop III.

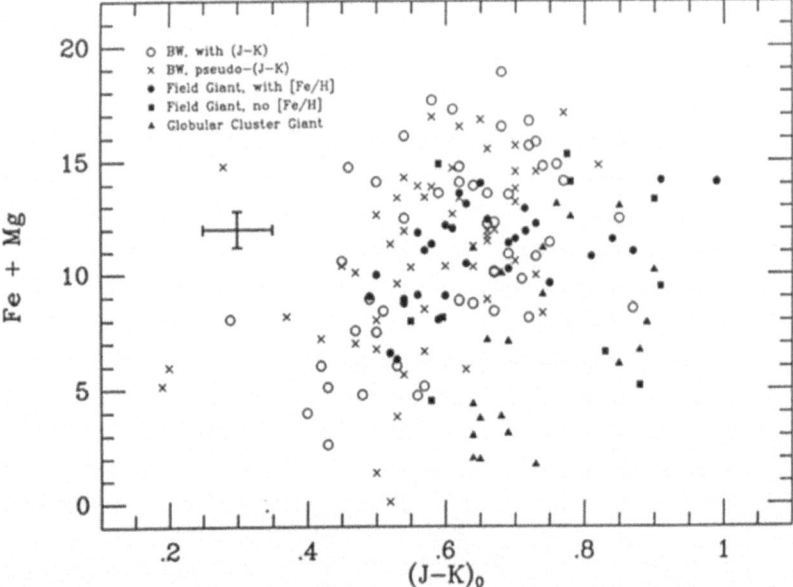

**Fig. 11.** The line strengths of bulge stars are compared with stars of known abundance, at constant (J-K) color (i.e. at constant effective temperature). Notice that bulge giants have clearly larger line strength at constant color than the solar neighborhood stars.

### 3.2. WINDS VS. DISSIPATION

The simplest additional complication to the One Zone model is significant removal of gas from the volume in consideration. Searle and Zinn (1978) and Pagel and Patchett (1975) explore the consequences of outflow of enriched gas. Gas can leave the volume for many reasons; in addition to the familiar SN-driven wind, gas could cool and dissipate toward the nucleus, leaving the volume under consideration. If the gas outflow is steady in time, the effective yield is decreased, but the functional exponential form is identical. If the outflow is catastrophic (for example, when the energy input due to SNe exceeds the binding energy) then star formation suddenly ceases, producing a sharp cutoff.

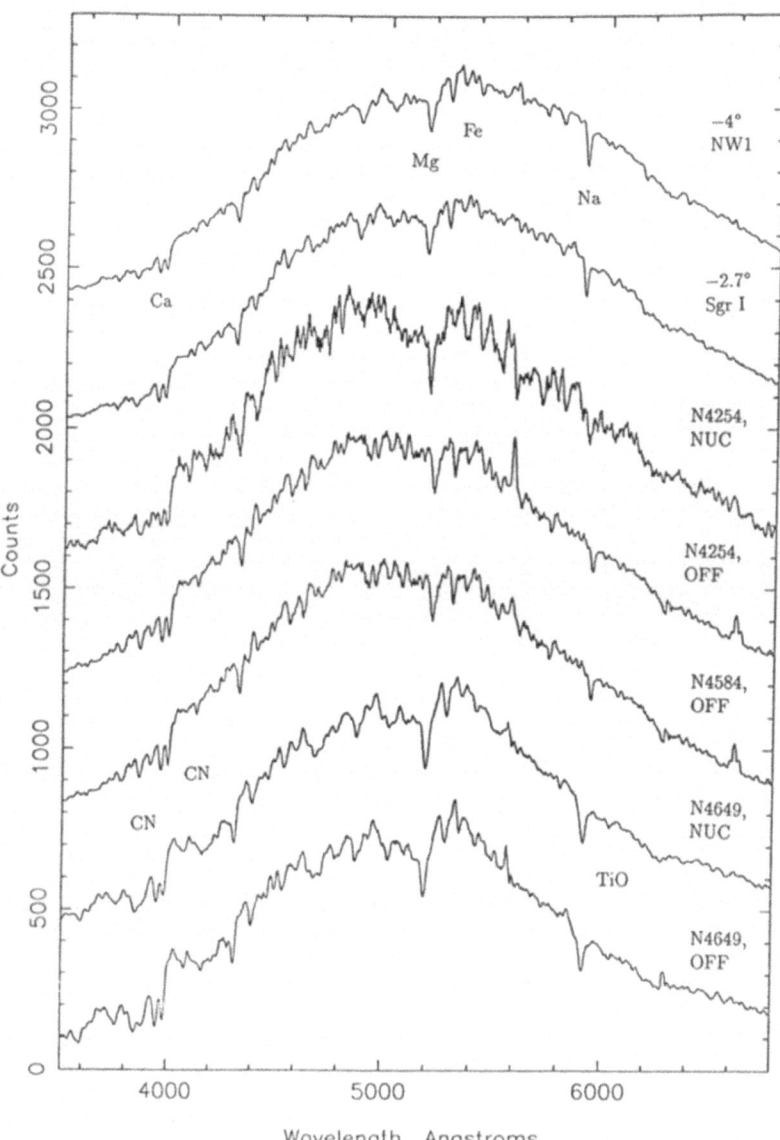

**Fig. 12.** The integrated light of bulge fields is compared with that of other bulges and galaxy nuclei. The galactic bulge fields have a smaller H and K break, and weaker CN features at 3850 and 4100Å. Overall, the bulge fields have weaker lines. Quantitative analysis finds no difference in line strength between SgrI and NW1, which are at different distances from the nucleus.

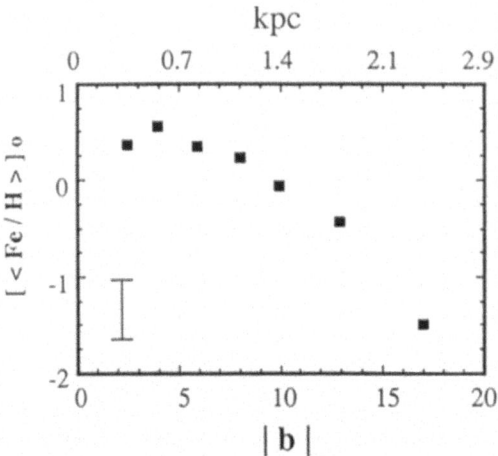

**Fig. 13.** Bulge abundance gradient derived from Washington photometry from Tyson's (1991) Ph.D. thesis. Notice that the abundance appears constant within $-8° = 1$ kpc. It is likely that the decline represents a transition to the halo population. The final bin at $17°$ is the mean of 3 stars.

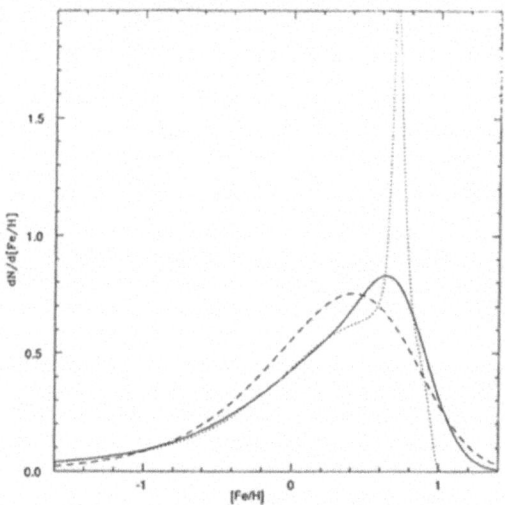

**Fig. 14** Strict "closed box" chemical evolution with mass return, from Zhao *et al.* 1991. *Dashed line:* simple model with a yield of twice solar; *dotted line:* theoretical abundance distribution including the large number of moderate metallicity stars formed from material returned by the first generation of intermediate mass stars. *solid line:* theoretical curve convolved with 0.2 dex gaussian errors, approximately what one measures with high dispersion spectroscopy. Notice that the peak is washed out, and the result resembles the Simple Model with a higher yield.

Franx and Illingworth (1991) show that color gradients in elliptical galaxies are correlated with local escape velocity. If this is generally the case, then winds play an important role in the chemical evolution. If the wind outflow is steady we can predict the abundance distribution will follow $n(z) = e^{-z/y}/y$, where y is the mean metallicity (which estimates the effective yield). The idea that color gradients are correlated with local escape velocity is a new idea, and would suggest that wind outflow rather than dissipation is responsible for color gradients in spheroidal components.

### 3.3. DISSIPATION

In the collective unconscious there is an expectation that abundances and kinematics must be connected, a picture of enriched gas dissipating into a disk and toward the nucleus. The ideas of Eggen *et al.* (1962) and Larson (1974) justifiably continue to foster the belief that there must be connections between kinematics and abundances in the stars of the bulge and the halo, despite observational indications that this is likely not the case (Carney *et al.* 1990).

The galactic bulge is rotating and is probably rotation supported; this is true as well for external bulges (Kormendy and Illingworth, 1982). In the case of the bulge, it appears that metal rich K giants may have a smaller dispersion than the metal poor stars (Rich, 1990b). The evidence mentioned in §3.1 contradicts earlier observations, and supports a population transition rather than abundance gradient. According to Larson's (1974) models, gradients are attributed to dissipation in the process of galaxy formation. It is virtually certain that the gas of the proto-bulge underwent substantial dissipation to produce a system of such high phase space density. Hernquist (1989) finds that gas can be delivered to the center in a dynamical time, a period too short for there to be a strong correlation with any enrichment.

In a collapse and enrichment scenario, metal rich gas flows toward the nucleus of a galaxy. Chemical evolution in the central region is strongly affected by infall of enriched material. This produces an abundance distribution that is skewed toward the metal rich end, peaking at the yield (cf. Mould, 1984). Such an abundance distribution is not observed at any location in the galactic bulge, whose abundance distributions are all skewed toward the metal poor end. One may speculate that enriched gas flows into the nuclei of galaxies, but while strong-lined, these populations do not appear dominated by late M giant light (refer back to Figure 12). The CO band, strong in late M giants, should increase dramatically toward the nucleus of M31; in fact, there is no change in CO between the bulge and nucleus (Rich and Mould, 1992).

Despite some evidence for a kinematics/abundance connection in the bulge (see also §4.2) it remains difficult to tell if this is due to the bulge/halo population mix, or the fossil of chemical/dynamical evolution. Large samples of stars with abundances and kinematics are required if we are to distinguish conclusively between winds and dissipation. We will also need to measure the density law in the bulge as a function of metallicity. Minnitti (1991) finds that rotation increases and velocity dispersion decreases for the most metal rich bulge stars. Whether this is because kinematics and abundance are connected in one population is uncertain. It is important to emphasize that the difference between two velocity dispersions is determined with an error of $\sigma/\sqrt{N}$, for two samples of size N. Measured dispersion differences (as a function of abundance) have been found to be ≈10 km/sec. If we wish a $5\sigma$ result for 100 km/sec velocity dispersions, we require each sample to have of order 2500 stars, five times the size of the largest sample to date. I have two programs in progress to achieve this level. In collaboration with S. van den Bergh, I am measuring proper motions for bulge giants in the critical −8° field, 1kpc from the nucleus. We expect to have some $10^4$ space motions and abundances in the next few years.

How can we belive that enriched gas dissipates toward the nucleus and forms stars, yet also believe that supernova-driven winds exhaust the spheroid of its gas at some point in its history? Is there any clear evidence for progressive chemical/dynamical evolution "collapse and spin-up" in *any* Milky Way stellar population?

### 3.4. DETAILED CHEMICAL ABUNDANCES: THE ENRICHMENT TIMESCALE

Supernova yields are now known to sufficient accuracy that we may separate the contribution from type I (core deflagration) and type II (core bounce) SNe. Massive star SNe produce mainly O and α−capture elements, while white dwarf (Type I) SNe make the majority of the Fe. The timescale for enrichment by I SNe is ≈ $10^7$ yrs or perhaps shorter, whereas the WD SNe enrichment timescale is likely $10^9$ yrs or longer. This difference is accepted as the reason that extremely metal poor stars have enhanced O abundances (see the review by Wheeler, Sneden and Truran 1990). Matteucci and Brocato (1990) report results of detailed models including both SN types and can reproduce the bulge abundance distribution function in Figure 10. They further propose that if bulges formed rapidly, stars well above the solar abundance should remain enhanced in O and other elements made in massive star SNe (see Figure 15).

In order to test this hypothesis, A. McWilliam and I have been measuring abundances of K giants in Baade's Window. We seek echelle spectra for 25 stars covering the full bulge abundance range, so that trends in elements can clearly be seen. Some results are already clear: Eu, a pure r-process

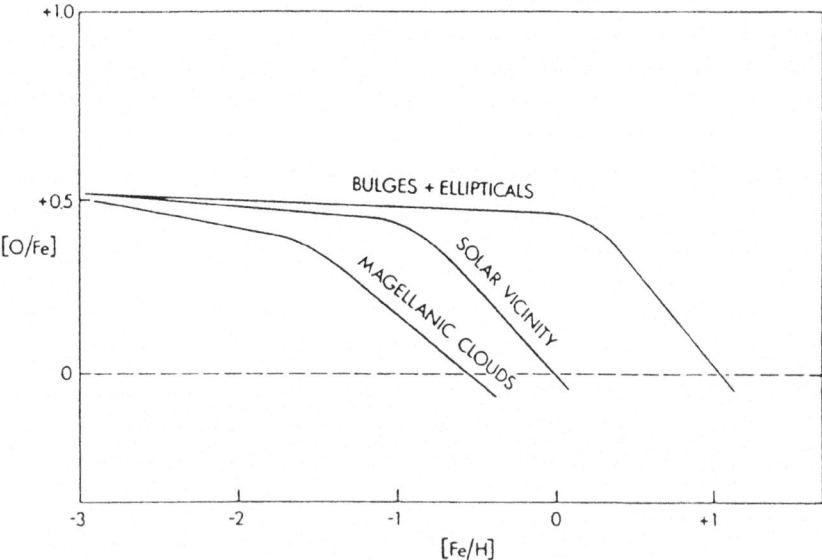

**Fig. 15.** Schematic trends of massive star nucleosynthesis products as a function of abundance, for several populations with different formation histories. The massive stars are presumed to produce O in SNe, although Ca, Si, and Eu (r-process elements) are also produced. Bulges and ellipticals form rapidly. Massive star SNe rapidly produce most of the metals, so the system has reached higher abundance before type II SNe become important. Echelle spectroscopy and detailed abundance measurement for bulge stars provide a test for these evolutionary scenarios. (Figure is from Matteucci and Brocato, 1990).

element, is greatly enhanced; Si and Ca are up a factor of two relative to the solar composition. It is thought that SNe of intermediate mass are the site of the r-process (Cowan *et al.* 1991), so we can conclude that the bulge enriched rapidly, but not violently (O is not dramatically enhanced). To reconcile this evidence for rapid enrichment with the luminous AGB, the bulge must have formed in a starburst taking place after the globular clusters. Baade (1963) summed up these results in a prescient flash of inspiration: "After the first generation of stars has been formed, we can hardly speak of a 'generation', because the enrichment takes place so soon, and there is probably very little time difference."

# 4. Kinematic Evolution

The search for a relation between kinematics and abundances of bulge stars continues to motivate much research, despite recent findings that the halo lacks any such connection (Carney *et al.* 1990). The formal calculation of relaxation time for the bulge finds it much larger than a Hubble time. However, recent thinking has dramatically altered the picture. Strong evolution, mixing, and dynamical heating can occur rapidly, *in a few dynamical times*. The most effective mechanisms appear to operate in bars, making it all the more important to establish whether or not the bulge has this geometry. If the bulge is superimposed on a halo population, we must interpret abundance/kinematic correlations as a superposition of populations. The bulge is a distinct experiment in galaxy formation, and may also give us the chance to study interesting mechanisms for secular dynamical evolution

## 4.1. ROTATION

If we consider the galactic bulge as being the region of high density confined within $\pm 6°$ of latitude, all of the stellar populations studied there rotate at the rate of $\approx 10$ km/sec per degree (Kinman *et al.* 1988; Catchpole, 1990; Menzies, 1990). At $b = -14°$, the rotation apparently drops from 100 km/sec to 70 km/sec (Harding, 1990). At the current time, only Minnitti *et al.* (1991) reports a possible correlation of abundances and rotation in the bulge. His fields lie near the edge of the 1 kpc inner bulge population, at $(l, b) = (8, 7)$. While it is attractive to propose that such a correlation supports a dissipative collapse model for the bulge, it is in principal equally consistent with a mixture of stellar populations. There is general consensus that the metal rich bulge stars have substantial rotation, of about 100 km/sec. The bulge OH/IR stars (cf. Winnberg *et al.* 1985, te Lintel Hekkert 1991) rotate slightly more rapidly. A strong bar potential has been proposed to explain the kinematics of the $10^8 M_\odot$ of molecular gas present in the bulge (Binney *et al.* 1991). In the bulge $^{12}CO$ reaches velocities of $\pm 200$ km/sec within $l = 2°$ and has a velocity field consistent with strongly noncircular motion.

## 4.2. VELOCITY DISPERSIONS

The early investigations by Feast (1986) found an approximate dispersion of 100 km/sec for bulge populations in the vicinity of Baade's Window at $b = -4°$. Workers have naturally sought to find correlations between velocity dispersion and abundance similar to what people thought was present in the stellar halo. Rich (1990b) reports that bulge giants with [Fe/H] < −0.3 have a dispersion of 120 km/sec, while those with $[Fe/H] > +0.3$ have a dispersion of 95 km/sec. Rich *et al.* (1992) find that bulge RR Lyraes (stars generally with [Fe/H] < −0.5 have $\sigma = 125$ km/sec. The correlations between abundance and dispersion are confirmed by Minnitti (1991) who finds that metal poor giants have larger dispersions than metal rich giants at several locations in the bulge. Independent confirmation of some effect comes from new proper motion data measured from Baade's original plates of the bulge in Baade's Window. Spaenhauer *et al.* 1991 find tantalizing results for correlations of abundances and kinematics in the bulge. They find that $\sigma_b$ decreases from 78 to 58 km/sec as the [Fe/H] increase above solar. Again, the result is at a $2\sigma$ level for a sample of 57 stars. It is curious that the $b$ velocity dispersion is smaller for those 57 stars with abundances than for the total sample of 429 stars. In considering the larger sample of 429 stars, Sadler, Terndrup, and Rich (1992) employ an Fe index measured from low dispersion spectra to find that both $\sigma_r$ and $\sigma_b$ decrease from $110 \pm 10$ to $90 \pm 10$ km/sec as [Fe/H] increases from −0.3 dex to > +0.3 dex. Figure 16 shows the distribution function of velocities for the largest samples of bulge kinematic probes. At sample sizes of a few hundred, dispersion differences become significant; the smaller vertical velocity dispersion in Figure 16 is significant. Abundance/kinematics correlations will require samples of similar size in each abundance group.

## 5. Evolution Scenarios for the Bulge

A conventional view of the bulge's history would have the dense proto-bulge as perhaps the first system in the Galaxy to collapse. Lee's (1991) chronology of the Galaxy based on RR Lyrae stars finds the galactic center RR Lyraes to be the oldest, 1 Gyr older than even the globular clusters. Rapid, even violent formation would be consistent with the Simple Model metallicity distribution and the r-process enhancements mentioned earlier. Perhaps the bulge is the site where the Galaxy behaved according to the ELS picture, with collapse, spin-up, and enrichment.

Yet the AGB stars appear to tell a different story (§2), that of a bulge *younger* than the globular clusters. The frontispiece of this book reminds us that some bulges even clearly possess early-type spectra (Fig. 17). We are challenged either to explain the very luminous AGB stars, or to explore alternative ideas in galaxy evolution. Recent developments compel us to consider tha latter course. Secular processes such as mergers, starbursts, and dynamical evolution—the astronomical equivalent of the gradual erosion of a mountain range—may affect galaxy evolution. If an accretion event delivers $10^9 M_\odot$ of gas to the nuclear region of a galaxy, a starburst in that gas would

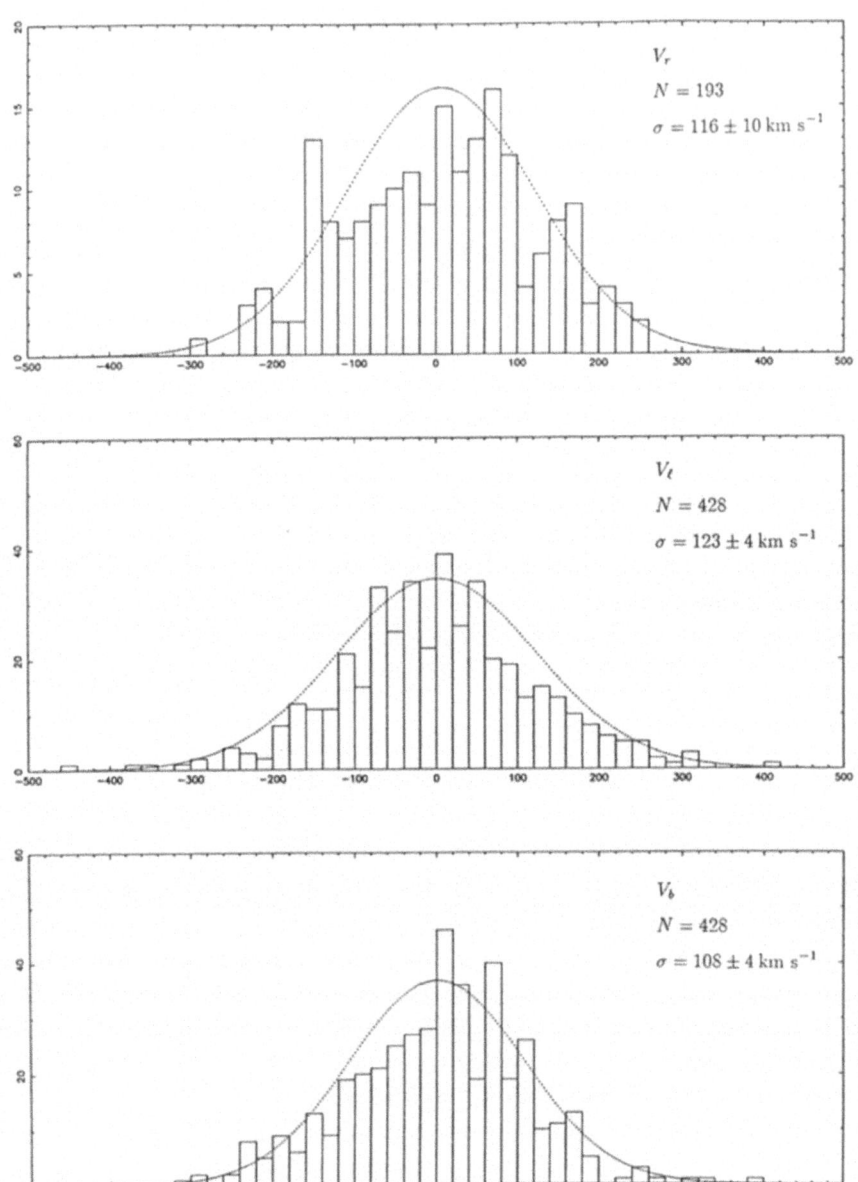

**Fig. 16.** Velocity distribution of bulge stars from two stellar samples. Radial velocities from Sharples *et al.* 1990 for Blanco's late M giants. Velocity dispersions in *l* and *b* are based on proper motions from Spaenhauer *et al.* 1992, and $R_0 = 8$kpc. The proper motion distributions satisfy a one-sided K-S test for being Gaussian. Notice that the *b* velocity dispersion (perpendicular to the plane) is significantly lower than the other dispersions.

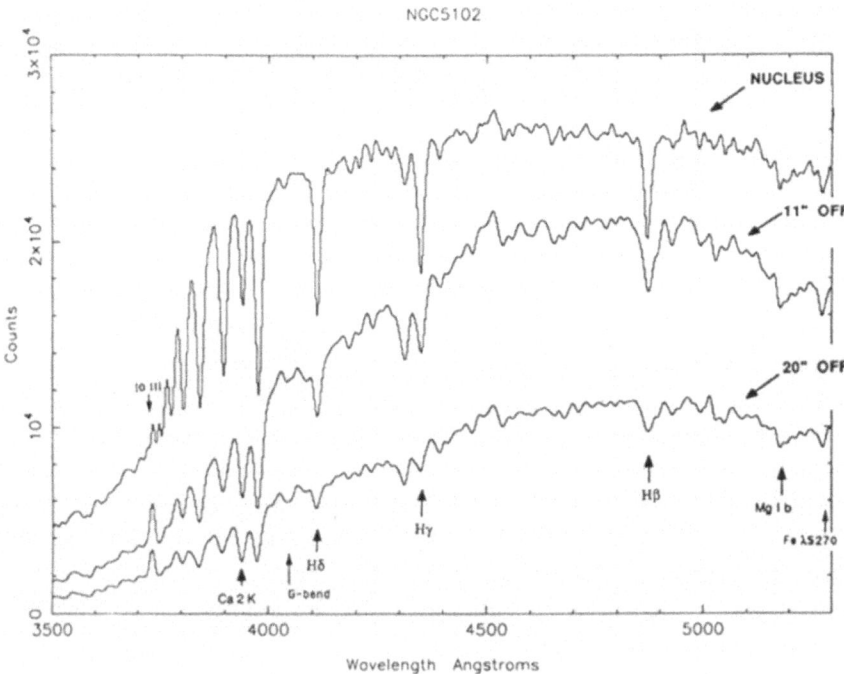

**Fig. 17.** Spectroscopy of the nucleus and bulge of NGC 5102, (Frontispiece) from Hibbard and Rich, 1990. Notice the clear persistance of Balmer lines even 30 arcsec ($\approx$ 600 pc) from the nucleus. This shows that the whole bulge of NGC 5102 was involved in a starburst approximately one Gyr ago. Perhaps starbursts may enable galaxies, like people, to build their bulges late in life.

be identical to one occurring in a proto-galaxy. Luminous galaxies observed by *IRAS* are sometimes found to have up to $10^{10} M_\odot$ of gas in their nuclear regions. Some of these galaxies are presently undergoing a starburst that, aged a few billion years, might eventually fade to a conventional red bulge.

As mentioned earlier, recent theoretical work by Hernquist (1989) supports the idea that gas can dissipate tremendously on the order a crossing time. It is attractive to consider such dissipation as an alternative means of forming bulges and nuclei of high phase space density well after the initial proto-galaxy collapse phase. The key new idea is that this scenario permits the build up of high phase-space density structures at any point in a galaxy's lifetime without demanding a corresponding metal enrichment for such structures.

One may query how a gas disk could thicken into a bulge. In fact, disks are unstable to bar formation, and such bars can thicken perpendicular to the plane in a few dyanmical times (Pfenniger and Norman, 1990; Combes *et al.* 1990; Raha *et al.* 1991). Sellwood's mechanism in which a bar bends and heats shows that substantial dynamical evolution can occur on timescales much less than the relaxation time, an absurdly long $10^{17}$ yr for the galactic bulge. When numerical simulations are run for such heated systems, the edge-on view distinctly resembles a boxy bulge. The secular acceleration of disk stars is a good example of a dynamical process that operates on timescales of order 10 crossing times. If these effects do occur, it is clear that straightforward stellar dynamics is inadequate to describe these critical evolutionary processes. Bar formation does not neatly fit in to the ELS picture; considerable evidence has been brought forth at this meeting that the galactic bulge has a bar structure (Blitz and Spergel 1991; Whitelock, these proceedings). Significantly, Blanco's star counts and Kent's IR photometry find a relatively thin bulge, less than 400 pc scaleheight. A relatively cool $z$ velocity dispersion, and rotation support, concur with the properties of a bar-like bulge.

Were the bulge to thicken after the completion of star formation, we would expect to see no abundance gradient in its stellar population. Recent surveys of the bulge using Washington photometry (Tyson, 1991) and low dispersion spectroscopy (Rich *et al.* 1992) suggest that the field population at 1000 pc is not distinctly different from that of Baade's Window. A cautionary note is that the decline in the number of late M giants (Blanco, 1988) is steeper than that of the integrated K light (Kent *et al.* 1991).

Even in the conventional scenario of bulge/spheroid formation, the dynamical timescale of $\approx 10^8$ yr is close to the SNe II enrichment timescale, but less than the (presumed) SN I enrichment timescale of $\approx 10^9$ yrs. This could lead to observable spatial gradients in the $\alpha$-capture and r-process elements, but not in the Fe-peak elements. If winds are important, we will have to consider gas cooling; if bar formation and heating plays a role, it is possible that the population will experience substantial dynamical mixing after all the elements have formed. Important questions about the evolution of the bulge can only be resolved when large numbers of high resolution spectra are obtained, permitting measurement of the detailed chemical abundances of various kinematic groups.

5.1. EVOLUTION BY MERGERS?

As considered above, the formation of a bulge due to a gas accretion

event and subsequent starburst appears plausible, leading to a wide age range
in bulge populations. An alternative proposal, first advanced by Tremaine,
Ostriker. and Spitzer (1975) is that bulges are built out of the accretion
of formed stellar systems. Schweizer and Seitzer (1988) find ripples and
shells (merger signatures) in disk galaxies. Rich (1989) argues against the
Galaxy's bulge forming via a series of mergers continuing to within a few
Gyr of the present day. I will briefly review these arguments here. As
Toth and Ostriker (1992) point out, infall of a formed stellar system would
have heated the disk far greater than what is observed. If objects like the
Magellanic Clouds accreted into the bulge over the last few Gyr, then these
low metallicity systems would leave swarms of luminous red carbon stars
(the smile on the Cheshire Cat). Extensive surveys by Blanco and Azzopardi
find no such luminous carbon stars (the luminosity function of Azzopardi's
C stars in Rich (1989) terminates a full magnitude below the He core flash
luminosity). Low mass galaxies lose their gas to winds before they become
sufficiently enriched to reach metallicities as high as those observed in the
bulge (cf. Arimoto and Yoshii, 1987).

## 6. Conclusions

The evolved stars of the galactic bulge challenge us either to explain
their high luminosities and mass loss rates, or accept a bulge that is younger
than the globular clusters. If the bulge is the oldest part of the Galaxy then
we must conclude that our understanding of metal rich stellar evolution
is deficient. If the bulge is younger than the globular clusters. we must
explain why this dense stellar subsystem formed so late. Less than 10% of
the luminous mass of the Milky Way would have formed at the time of the
globular clusters. This may sound like a surprising conclusion. but there
are well known examples of local group systems with such a peculiar star
formation history. The Magellanic Clouds and the Carina dwarf spheroidal
(cf. Mighell 1991, Mould and Aaronson 1983) are two prime examples. In
the case of Carina, 90% of the luminous mass is in a 7 Gyr old population,
while an underlying 10+Gyr old (globular cluster-like) population accounts
for only 10% of the luminous mass. Searches for primeval galaxies at high
redshift have been disappointing. Objects expected to be bright at Lyman-
alpha have not turned up. Perhaps the formation of galaxies (even their
high phase-space density components) takes place over an extended time.
as Katz and Gunn (1992) have suggested. So we return to a reminder that
bulges can in fact be young; Figure 17 shows the spectrum of the galaxy in
the frontispiece, NGC 5102. with Balmer lines visible over the whole bulge.

While many of these issues remain unsettled, it is clear that the galactic
bulge may offer the best analog to elliptical galaxy evolution. Study of the

dynamics of gas and stars, as well as the detailed chemistry of the stars themselves, will doubtless deepen our understanding of how ellipticals and bulges form.

The author acknowledges valuable discussions with J. Applegate and D. Spergel. The author is also grateful for F. Feliciano's careful preparation of the camera-ready manuscript, and for the assistance of Sonya Umar in preparation of the figures.

# References

Arimoto, N. and Yoshii, Y. 1987, *Astr. Ap.*, **173**, 23

Armandroff, T.E. 1989, AJ, **97**, 375

Baade, W. 1963 in *The Evolution of Galaxies and Stellar Populations* (Harvard U. Press: Cambridge) p.279

Bahcall, J.N., Schmidt, M., and Soneira, R.M. 1983, *Ap.J.*, **265**, 730

Barbuy, B., and Grenon,M. 1990 in *Bulges of Galaxies*, B. Jarvis and D. Terndrup, eds. p. 47

Bally, J., Stark, A.A, Wilson, R.W., and Henkel, C. 1988, *Ap.J.*, **324**, 223

Binney, J. *et al.* 1991, *M.N.R.A.S.*, **252**, 210

Blanco, B. 1984, *A.J.*, **89**, 1836

Blanco, V. 1988, *A.J*, **95**, 1400

Blanco, V.M. 1992, *A.J.*, in press

Blitz, L., and Spergel, D.N. 1991, *Ap.J.*, **379**, 631

Butcher, H.R., and Oemler, A. 1978, *Ap.J.*, **219**, 18

Carney, B.W., Latham, D.W., and Laird, J.B. 1990, in *Galactic Bulges*, B. Jarvis, D. Terndrup, eds. p. 127

Catchpole, R.M. 1990, in *Galactic Bulges*, B. Jarvis, D. Terndrup, eds. p. 111

Combes, F., Debbasch, F., Friedli, D. and Pfenniger, D. 1990, *Astr. Ap.*, **233**, 82

Cowan, J.J., Thielemann, F-K, and Truran, J.W. 1991, 208, 26

Davies, R.L., Frogel, J.A., and Terndrup, D.M. 1991, *A.J.*, **102**, 1729

Eggen, O.J., Lynden Bell, D., and Sandage, A.R., *Ap.J.*, **136**, 748

Feast, M.W. 1963, *M.N.R.A.S.*, **125** 27

Feast, M.W. 1986, in *Light and Dark Matter*, ed. F. Israel (Dordrecht: Reidel), p. 339

Franx, M., and Illingworth, G. 1990, *Ap.J. (Lett.)*, **359** L41

Freedman, W. 1991 in, *Stellar Populations, IAU Symp. 149*, A. Renzini and B. Barbuy, eds. Kluwer

Frogel, J.A. 1988, *Ann. Rev. Astr. Ap.*, **26**, 51

Frogel, J.A., and Whitford, A.E. 1982, *Ap.J. (Lett.)*, **259** L7

Frogel, J.A., and Whitford, A.E. 1987, *Ap.J.*, **320**, 199

Frogel, J.A., Terndrup, D.M., Blanco, V.M. and Whitford, A.E. 1990, *Ap.J.*, **357**, 453

Geisler, D. and Friel, E.D. 1992, *Ap.J.* in press

Green, E., and Norris, J. 1990, *Ap.J. (Lett.)*, **353**, L17

Habing, H.J. *et al.* 1985, *Astr. Ap.*, **152**, L1

Harding, P. 1990, in *Galactic Bulges*, B. Jarvis, D. Terndrup, eds. p. 105

Harmon, R.T., and Gilmore, G. 1988, *M.N.R.A.S.*, **235**, 1025

Hernquist, L. 1989 in 14th Texas Symp., *Annals of NY Acad. Sci.*, **571**, 190

Katz, J., and Gunn, J. 1992, *Ap.J.*, in press

Kent, S.M. 1989, *A.J.*, **97**, 1614

Kent, S.M., Dame, T.M., and Fazio, G. 1991, *Ap.J.*, **378**, 496

Kinman, T.D., Feast, M.W., and Lasher, B.M. 1988, *A.J.*, **95**, 804

Kormendy, J., and Illingworth, G. 1982, *Ap.J.*, **256**, 456

Maeder, A., and Meynet, G. 1988, *Astr. Ap. Supp.*, **76**, 411

Larson, R. 1974, *M.N.R.A.S.*, **166**, 585

Lee, Y.W. 1992 in , *Stellar Populations, IAU Symp. 149*, A. Renzini and B. Barbuy, eds. (Kluwer:Dordrecht)

Matteuci, F., and Brocato, E. *Ap.J.*, **365**, 539

Mighell, K.J. 1990, *Astr. Ap. Supp. Ser.*, **82**, 1

McWilliam, A., and Rich, R.M. 1992, in preparation

Menzies, J.W. 1990, in *Galactic Bulges*, B. Jarvis, D. Terndrup, eds. p. 115

Minitti, D. 1991 in *Stellar Populations, IAU Symp. 149*, A. Renzini and B. Barbuy, eds., Kluwer

Mould, J., and Aaronson, A. 1983, *Ap.J.*, **273**, 530

Norris, J. 1986, *Ap.J.*, **61**, 667

Norris, J., and Ryan, S. G. 1989, *Ap.J.*, **336**, L17

Oort, J. 1977, *Ann. Rev. Astr. Ap.*, **15**, 295

Ortolani, S., Barbuy, B., and Bica, E. 1990, *Astr. Ap.*, **236**, 362

Ortolani, S. 1992, *Astr. Ap.* in press

Pagel, B.E.G., and Patchett, B.E. 1975, *M.N.R.A.S.*, **172** 13

Pfenniger, D., and Norman, C. 1990, *Ap.J.*, **363**, 391

Raha *et al.* 1991, *Nature*, **352**, 411

Renzini, A., and Greggio, L. 1990, in *Bulges of Galaxies*, B. Jarvis and D. Terndrup, eds. p. 47

Renzini, A., and Buzzoni, A. 1986, in *Spectral Evolution of Galaxies*, ed. C. Chiosi, A. Renzini (Dordrecht: Reidel), p. 135

Rich, R.M. 1989, in *The Center of the Galaxy, IAU Symp. 136*, ed. M. Morris (Dordrecht: Kluwer), p. 63

Rich, R.M. and Mould, J.R. 1991, *A.J.*, **101**, 1286

Rich, R.M. 1985, *Mem.S.A.It.*, **56**, 23

Rich, R.M. 1988, *A.J.*, **95**, 828

Rich, R.M. 1990a, in *Bulges of Galaxies*, B. Jarvis and D. Terndrup, eds. p. 65

Rich, R.M. 1990b, *Ap.J.*, **362**, 604

Rich, R.M. 1991 in *Stellar Populations, IAU Symposium 149*, A. Renzini, B. Barbuy eds. Dordrecht: Kluwer

Rich, R.M., Terndrup, D., and Sadler, E. 1992, in preparation

Rich, R.M., Mould, J., and Graham, J. 1992, in preparation

Rich, R.M., and Mould, J. 1992, in preparation

Saha, A. 1985, *Ap.J.*, **289**, 310

Sadler, E., Terndrup, D., and Rich, M. 1992, in preparation

Schuster, W.J., and Nissen, P.E. 1989, *Astr. Ap.*, **222**, 69

Schweizer, F. 1982, *Ap.J.*, **252**, 643

Schweizer, F., and Seitzer, P. 1988, *Ap.J.*, **328**, 88

Searle, L., and Sargent, W.L.W. 1972, *Ap.J.*, **173**, 25

Searle, L., and Zinn, R. 1978, *Ap.J.*, **225**, 357

Sharples, R., Walker, A., and Cropper, M. 1990, *M.N.R.A.S.*, **246**, 54

Spaenhauer, A., Jones, B., and Whitford, A.E. 1991. *A.J.*, **103**, 297

Suntzeff, N. *et al.* 1991, *Ap.J.*, **367**, 568

te Lintel Hekkert, P. 1990, Ph.D. Thesis, Leiden University

Terndrup, D.M. 1988, *AJ*, **96**, 884

Terndrup, D.M. *et al.* 1990, **357**, 453

Toth, J., and Ostriker, J. 1992, *Ap.J.*, in press

Tremaine, S.D., Ostriker, J.P., and Spitzer, L., Jr. 1975, *Ap.J.*, **196**, 407

Tyson, N.D. 1991, Ph.D. Thesis, Columbia University

van der Veen, W., and Habing, H.J. 1990, *Astr. Ap.*, **231**, 404

vanden Berg, P.A., Bolte, M., and Stetson, P. 1990, *A.J.*, **100**, 445

Walker, A., and Terndrup, D. 1991, **Ap.J.**, **378**, 119

Whitelock, P.A. 1990 in *Confrontation Between Stellar Pulsations and Evolution*, eds. C. Cacciari and G. Clementini (ASP: San Francisco), p. 365

Whitelock, P.A. *et al.* 1991, *M.N.R.A.S.*, **248**, 276
Winnberg, A. *et al.* 1985, *Ap.J. (Lett.)*, **291**, L45
Wood, P.R., and Bessell, M.S. 1983, *Ap.J.*, **265**, 748
Yamauchi, S. *et al.* 1990, *Ap.J.*, **365**, 532
Zhao, H., Applegate, J., and Rich, R.M. 1992, in preparation

# THE SHAPE OF THE GALAXY

DAVID N. SPERGEL

*Princeton University Observatory,*
*Princeton, NJ 08544-0001 USA*

**Abstract.** There is growing evidence that our Galaxy is a complex triaxial system. In this article, I begin by reviewing orbital structure in a triaxial galaxy and then review the evidence for the bulge being barred and the evidence about whether the Galactic spheroid is triaxial. I conclude by discussing the implications of triaxiality for galaxy formation and evolution.

## 1. Introduction

What is the shape of the Galaxy? In this talk, I will review several recent papers that suggest that the Galaxy is not axisymmetric. Since there is not yet a successful synthesis, the goal of this talk is to raise questions rather than to provide a comprehensive synthesis.

We will begin by reviewing the major stellar orbit families. The different stellar components will populate these different orbital families. In section 3, we will discuss several recent papers that suggest that the bulge of the Galaxy is bar-shaped. In section 4, we will focus on the Galaxy outside the inner 4 kiloparsecs and discuss the evidence for a triaxial spheroid and discuss whether the Galaxy has a lopsided gaseous disk.

In the concluding section of the talk, we will discuss the implications of triaxiality for galactic structure and evolution. As Mike Rich is fond of emphasizing, galactic astronomers are archaeologists. Stellar and gas kinematics and metal abundances are the artifacts that reflect the formation and evolution of our Galaxy. The existence of several triaxial components with different metallicity distribution hints at a complex galactic history. The existence of hundreds of millions of solar masses of gas in the Galactic bulge implies that it is a history not yet fully written: our Galaxy should be viewed as a dynamic system.

## 2. Orbits

In this section, we will review the major orbital families in a triaxial galactic potential. For a more comprehensive discussion of galactic orbits, see chapter 3 of Binney and Tremaine (1987), Contopoulus and Grosbol (1989) and Pfenniger and Freidli (1991).

We will begin by considering orbits in an axisymmetric oblate galaxy, which is symmetric about one short axis. We will then consider orbits in a

*L. Blitz (ed.), The Center, Bulge, and Disk of the Milky Way, 77–102.*

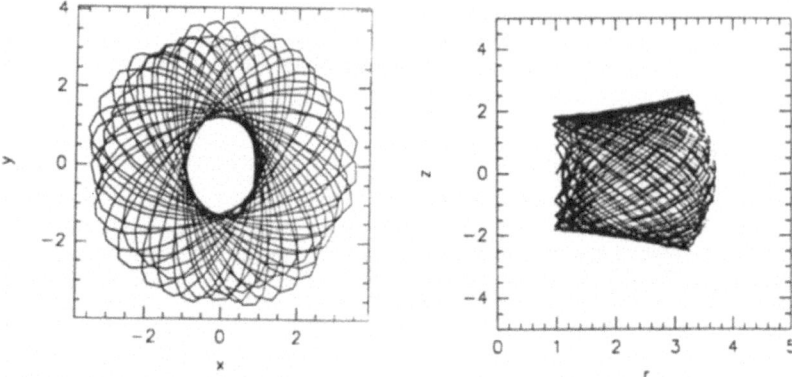

Fig. 1. A short axis orbit in an axisymmetric potential. Figure 1a show the path of the orbit in the x-y plane. Figure 1b shows the path of the orbit in the r-z plane. The Sun's orbit is a short axis loop orbit.

triaxial galaxy, which is reflection symmetric $(x \rightarrow -x, y \rightarrow -y, z \rightarrow -z)$ around a short, intermediate and long axis. In our own Galaxy, where we are near the plane of the disk, the minor axis is the short axis of the Galaxy and the major axis lies in the plane defined by the intermediate and major axis.

The simplest set of orbits are the circular orbits in a Keplerian potential. If we perturb a particle on a closed circular orbit, then it will oscillate around the circular orbit. In this potential, its radial epicyclic frequency equals its orbital frequency, thus the epicyclic motion produces a closed elliptical orbit. However, in a typical non-Keplerian potential, the radial and azimuthal frequencies are not equal, and the particle will move on a rosette orbit (see figure 1). This rosette orbit can be viewed as a member of an orbital family mothered by closed circular orbit. For example, the orbit of a star like our Sun is a member of the short axis loop family— the Sun's orbit consists of epicyclic oscillations around the closed loop orbit.

In an axisymmetric galaxy, there is only one important set of parent orbits: the closed short-axis loop orbits. These circular orbits are located in the plane of the galaxy and move around the short axis. They are specified entirely by $L_z$, the angular momentum around the short axis. Understanding closed orbits is an important step to understanding galactic structure and gas kinematics. Since gas clouds on orbits that self-intersect will collide with other clouds and lose energy, the gas eventually settles down into closed non-intersecting orbits. Thus, the gas streamlines are similar to closed loop orbits.

The orbital structure of a triaxial galaxy is significantly richer than an axisymmetric galaxy. In a non-rotating triaxial galaxy, the closed short axis loop orbit still exists, however, it is now elliptical rather than circular with its major axis perpendicular to the major axis of the potential (See Figure 2). All of the short-axis loop orbits spawned by the parent closed orbit have the same orientation. Thus, the quadrupole moment of the stars on loop orbits oppose the quadrupole of the underlying potential. (Note that in figure 2, the orbit is extended in a direction perpendicular to the long axis of the potential). This implies that it is impossible to construct a non-rotating triaxial galaxy in which all of the mass moves on loop orbits: an isolated elliptical disk cannot be an equilibrium system.

In a triaxial galaxy, a new orbit family arises: the box orbit. The mother of this family is a closed orbit along the long axis of the potential. Its offspring are orbits that oscillate in $y$ and $z$ around this closed orbit. Note that unlike the loop orbit, the long axis of these box orbits is aligned with the long axis of the triaxial potential. These box orbits are needed to make a self-consistent triaxial galaxy and are the backbone of any triaxial galaxy model. While most of the stars in a bar are part of this orbital family, gas in a non-rotating potential cannot settle down on these box orbits, since they are self-intersecting. Any gas on these orbits would self-intersect, shock and rapidly spiral inwards. In a galaxy that is only mildly triaxial, only a small fraction of the phase space, the region near the long axis, is occupied by this orbital family. As trixiality increases, this orbital family gradually grows at the expense of the short-axis loop orbits.

Yet another new closed loop orbit arises in a triaxial potential: a closed long-axis orbit. This orbit is confined to the y-z plane and orbits around the long axis of the galactic potential. A majority of the stars near the galactic minor axis can be part of this orbital family. If there is a preferred sense of rotation for these orbits, the galaxy may exhibit minor axis rotation. Gas can settle to the closed long-axis loop orbit and form a polar ring. Since this orbital family does not exist in an axisymmetric potential, the detection of a polar ring or minor axis rotation can be viewed as definitive evidence for non- axisymmetry in a galaxy.

In a non-rotating (or slowly rotating) galaxy, the existence of several orbital families allows the existence of three dynamically distinct components: (a) an elliptical disk of gas and stars whose major axis is perpendicular to the bar; (b) a triaxial stellar component consisting of stars on box orbits (an observer moving on a loop orbit would observe that this component has a large asymmetric drift); and (c) a stellar disk-like component orientated perpendicular to the galactic plane. As the triaxiality increases, the new orbital families (b and c) become increasingly important. It is tempting to associate the stellar disk (both thin and thick) with the short-axis loop orbits and the stellar spheroid with its nearly radial orbits with the box orbit family. These dynamically distinct components may have different formation and evolutionary histories.

In addition to the major orbital families, there are also sub-families

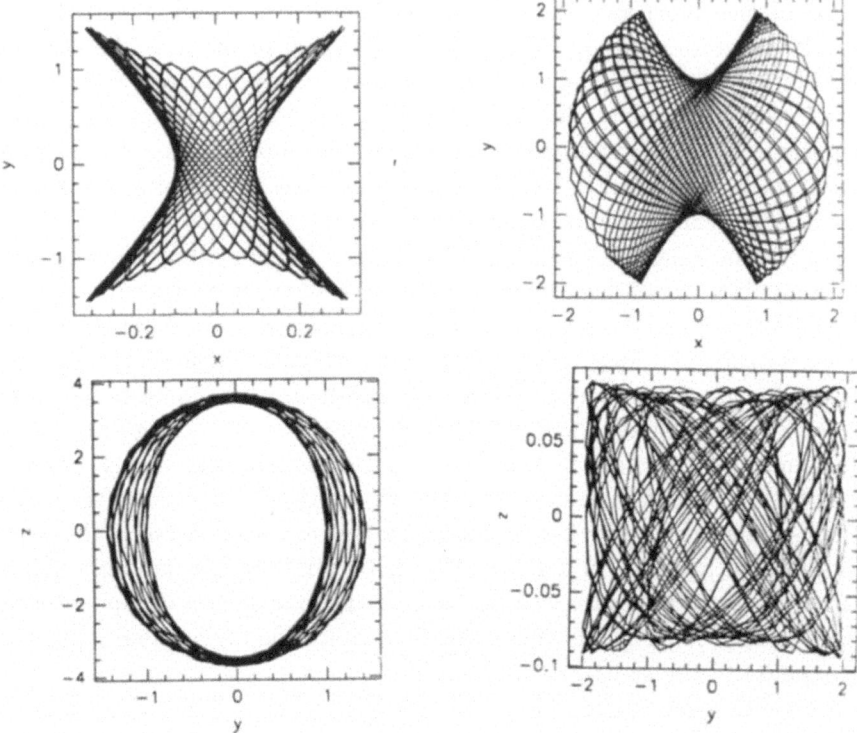

Fig. 2. The other two major orbital families in a non-rotating triaxial potential. The upper row of figures shows the path of the orbits projected onto the x-y plane. The lower row of figures shows the path of the orbit projected onto the y-z plane. The major axis of the potential lies along the x axis and the minor axis of the potential lies along the z axis.

spawned by resonant orbits. If the radial or vertical oscillation frequency is a rational product of the azimuthal frequency, then the star's orbit can eventually close on itself. This special orbit is called a closed resonant orbit. Schwarzschild and his collaborators have studied the various resonant orbits in non-rotating potential and have identified closed "banana", "fish" and "pretzel" orbits and their subfamilies (Miralda-Escude and Schwarzschild 1989, Lees and Schwarzschild 1991). It is not clear whether a significant fraction of the stars are part of these subfamilies.

The orbital structure in a rotating triaxial potential is even richer. At a given energy, there can now be more than one stable family of loop orbits in the galactic plane and these orbits can intersect. Contopoulus and Mertzanides (1977) have classified the orbits in a rotating bar potential. We will discuss these orbits in more detail when we review the Binney *et al.*

(1991) dynamical model for the gas in the Galactic center.

## 3. The Galactic Bulge

What is the shape of the bulge? Most astronomers describe the Bulge as an oblate rotator (see, *e.g.* Gilmore, King and Searle 1990), an axisymmetric system supported by stellar rotation and composed only of older stars formed during the genesis of the Galaxy (see, *e.g.* Frogel 1988). However, there has been growing evidence that Bulge may be a rapidly rotating triaxial bar, in which star formation has not ceased.

Infrared observations of the Galactic Bulge show that it is not a spherical system. Counts of M Giants along the major and minor axis show that the bulge is a flattened system (Blanco & Terndrup 1989). Similar flattening is seen in the distribution of OH/IR stars (Harmon and Gilmore 1988). Kent *et al.* (1991) use the 2.4 $\mu$m photometry of the bulge observed with the Spacelab Infrared Telescope (IRT) to reveal a flattened centrally concentrated stellar system that is well-fit by a exponential light distribution with a scale length of 400 parsecs along its minor axis and an axis ratio, $b/a$ of 0.61. COBE's beautiful multi-color image of the Galaxy clearly shows that the bulge is flattened and "peanut-shaped".

There are several independent pieces of evidence that suggest that the bulge of our Galaxy is triaxial and should be more probably viewed as a bar. In section 3.2, I will discuss the complex non-circular motion of the neutral hydrogen and molecular gas in the inner Galaxy that is likely to be a signature of a bar. Infrared photomotery, discussed in section 3.3, and star counts, discussed in section 3.4, provide direct evidence that the Bulge is triaxial. Stellar kinematics, discussed in section 3.5, do not show unambiguous evidence of either triaxiality or axisymmetry. However, future kinematical observations may provide a definitive test of triaxiality. In section 3.6, I discuss some of the implications of the Galactic bar for our understanding of the dynamics of bars. The effects of the bar on star formation and the chemical evolution of the Galaxy is discussed in section 3.7.

### 3.1. KINEMATICS OF THE GALACTIC CENTER GAS

Observations of neutral hydrogen provided the first hint of a non-axisymmetric Bulge. The distribution of gas in the $(l, v)$ diagram differs significantly from what is expected in an axisymmetric galaxy. de Vaucouleurs (1964), noting the similarity between the amplitude of the non-circular motions in our Galaxy and that observed in barred galaxies first suggested that the Milky Way is a barred galaxy. Gerhard and Vietri (1986) suggest that the non-Keplerian fall-off seen in the terminal velocity curve of the inner Galaxy is a signature of triaxiality. Various authors (Yuan 1984, van Albada 1985,

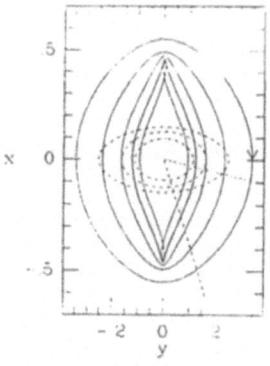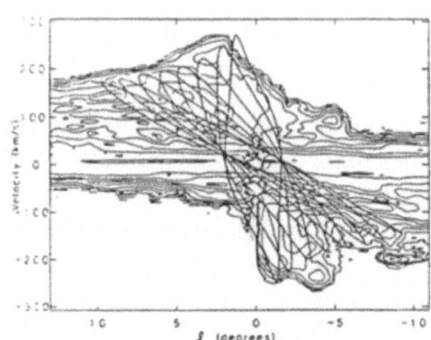

Fig. 3. (a) Closed prograde orbits in a barred rotating potential. The solid line shows the $x_1$ orbital family and the dashed line shows the $x_2$ orbital family. (b) (l-v) traces of $x_1$ orbits in the model potential, for an observer near the long axis of the bar, superposed onto intensity contours of 21 cm emission at $|b| < 0.5°$ from Burton and Liszt (1978). Figure from Binney et al. (1991).

Mulder and Liem 1986) suggest that the 3 kiloparsec expanding arm and the asymmetric distribution of gas in the Galactic center are due to the gas in the center of the Milky Way responding to forcing by a bar.

Recently, in an elegant paper, Binney et al. (1991) have constructed a coherent dynamical model for the HI, CO and CS emission in the inner 10° of the Galaxy. In their model, the Galaxy has a bar that extends out at least 1.2 kiloparsecs with corotation at 2.4 kiloparsecs. The bar in their model has the same orientation as that suggested in our simultaneous independent analysis of the photometry (section 3.3) and by star counts (section 3.4).

In a galaxy with a rotating bar, the prograde short-axis loop orbit family bifurcates into two distinct orbital families (Contopoulos and Mertzanides 1977). Deep inside the bar lie the $x_2$ orbits, positioned along the major axis of the bar. Further out, these orbits co-exist with the $x_1$ orbits whose long axis lies perpendicular to the major axis of the density distribution. At large radii, the $x_1$ orbits are non-intersecting; while at small radii, they are self-intersecting.

Binney et al. (1991) suggest that the HI initially follows the $x_1$ orbits. Eventually, the gas will drift down to the cusped $x_1$ orbit, where it will self-intersect and shock. They suggest that molecular gas forms in these shocks and then plunges onto the lower energy $x_2$ orbits. In their model, they quite convincingly fit the HI distribution in the $l-v$ diagram with non- intersecting $x_1$ orbits and the CO distribution with the cusped $x_1$ orbit. The cusp $x_1$ orbit projected into the $l-v$ diagram appears as a parallelogram, reproducing the gas distribution seen in the Galactic plane in the Bell Labs survey of Bally et al. (1987, 1988). Binney et al. (1991) suggest that the strong left-right asymmetry of the parallelogram can be partially understood as a perspective effect due to the nearer side of the bar lying at positive longitude.

Binney *et al.* (1991) note that their model is consistent with several other features in the gas distribution. In a barred galaxy, there are no stable, nearly circular orbits near corotation. This may explain the relative absence of HI and CO emission between 1.5 and 3.5 kiloparsecs. They note that hydrodynamical simulations find a build-up of gas near the outer Lindblad resonance (OLR). In their model, the OLR is located near the 3.5 kiloparsec molecular ring.

Yuan (1984) has shown that the expanding 3.5 kiloparsec arm can be understood as the gas response to a bar with roughly the same orientation and rotation rate as suggested in Binney *et al.* (1991).

## 3.2. PHOTOMETRIC EVIDENCE

If the stellar distribution is axisymmetric, then there should be no systematic difference between its appearance at positive and negative longitudes at infrared wavelengths, where the light is dominated by evolved stars and the extinction is neglible. However, if the Galaxy is barred and the Sun is not near either the major or minor axis of the bar, then we would expect systematic differences in both photometry and star-counts. Do the infrared maps of the Galactic center shows such asymmetries?

In Blitz and Spergel (1991b), we analyzed the Matsumoto *et al.* (1982) map of the 2.4 $\mu$m emission within 12° of the Galactic center in longitude and 10° of the center in latitude. We re-gridded the Matsumoto *et al.* (1982) data in 1° bins by interpolating from their published map and then looked for asymmetries in the light distribution. We restricted our analysis to $|b| > 3°$, to avoid disk emission. As is graphically seen in the three-color COBE photographic representation (Hauser *et al.* 1990) of the Galactic near-IR emission which shows a significantly reddened disk, but an unreddened bulge, the region outside $|b| > 3°$ is not significantly affected by dust.

Figure 4 shows systematic excess emission at positive longitude over the emission at negative longitudes. We refer for simplicity to longitudes between $l = 180°$ and $l = 360°$ negative longitudes. Note that there is a systematic excess of IR emission at positive longitudes over negative longitudes. In an axisymmetric galaxy, instrumental noise and local fluctuations in the star and gas distribution would produce just random positive and negative numbers in Figure 4. The excess of emission at positive longitudes shows that the distribution of stars is either asymmetric with respect to the Galactic center, a dynamically unstable situation, or that the distribution of stars is in the shape of a bar with the part of the long axis nearest the Sun occurring at positive longitudes.

If the bulge *is* bar-like with the long axis on the near side of the Galactic

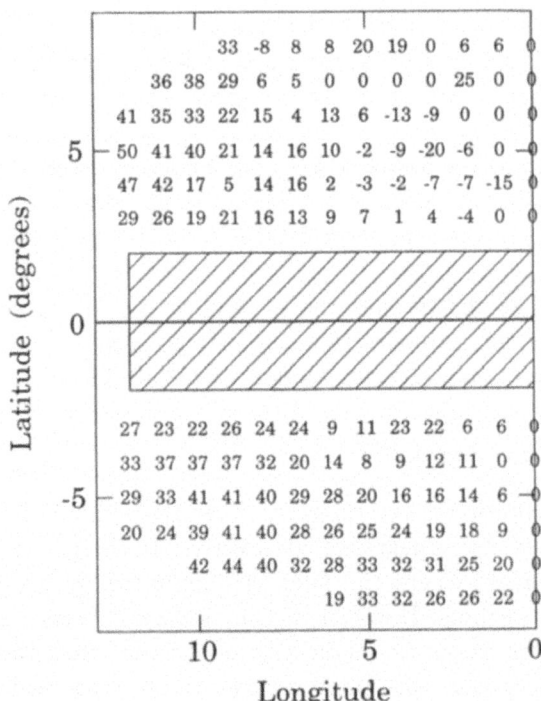

Fig. 4. The percentage difference, $\Delta I(l, b)$ between the negative and positive longitudes for the Matsumoto *et al.* (1982) Galactic center data. The values are plotted every degree; the resolution of the data is 0.°5. The estimated 1 $\sigma$ uncertainty in the map is about 5% - 7 % . The striped region along the Galactic plane represents the disk emission excluded from the analysis. (From Blitz and Spergel 1991b)

center in the first quadrant, then we would expect that the latitude distribution at positive longitudes should have a systematically greater angular scale height than at the corresponding negative longitudes. Figure 5 shows that the angular scale height is indeed systematically higher at positive longitudes.

## 3.3. STAR COUNTS IN THE BULGE

Star counts can provide unambiguous evidence for triaxiality. If the stellar population is triaxial, then the stars on the near side of the bar should

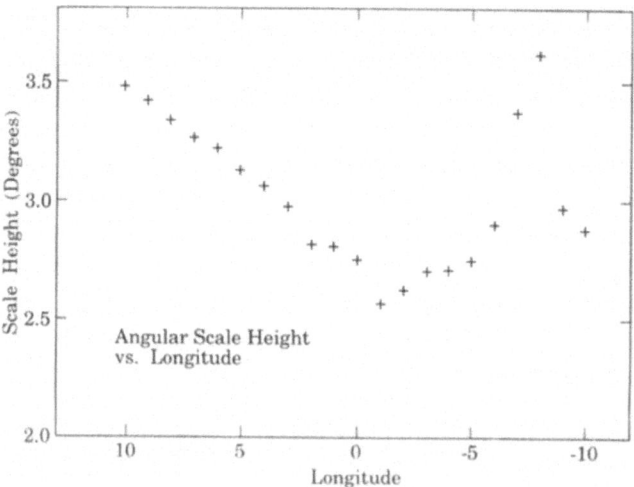

Fig. 5. Plot of the scale height as a function of Galactic longitude. The scale height was computed assuming an exponential surface brightness profile. Note that the scale heights are systematically higher at positive longitude. The two points that deviate from the trend are due to two bright spots on the Matsumoto *et al.* maps. An analysis of the Palomar plates finds globular clusters at these longitudes.

appear more luminous and be more numerous in any magnitude limited sample. Whitelock in her article in this book presents exciting new evidence for an asymmetry in the Mira counts, which is consistent with a bar with the same orientation as proposed in Blitz and Spergel (1991b) and by Binney *et al.* (1991). Contemporary with her work, Nakada *et al.* (1991) presented evidence of asymmetries in the IRAS bulge counts and in the distribution of planetary nebulae.

Nakada *et al.* (1991) analyzed the distribution of IRAS bulge stars and found a striking asymmetry. They determined the luminosity distribution for the IRAS sources with color $0 < \log(F_{25}/F_{12}) < 0.1$ and in the region bounded by $|b| < 3°$ and $|l| < 10°$. This criterion selected bulge stars with a dust shell temperature of $\sim 280$ K. They found that there were more sources at positive longitudes and that these sources were systematically brighter. Both results are statistically significant. They show that the luminosity distribution of sources with $l > 0$ is shifted relative to the luminosity of sources with $l < 0$. This shift is consistent with the sources at $l > 0$ having an average distance of 7.5 kpc and the sources at $l < 0$ having an average distance of 8.5 kpc, an additional important piece of evidence for triaxiality. Their counts also hint that the Galactic Bulge is tilted with respect to the disk of the Galaxy.

In interpreting star counts, it is important to remember that different stellar populations respond differently to the bar. If we imagine that the bar formed out of a preexisting stellar disk, then most of the older (pre-bar formation) stars should be primarily on $x_2$ orbits, which support the bar. Star counts of this population should show that the star counts have the same asymmetry as the mass distribution. Stars that form out of a gas disk that sitting in the bar potential will be primarily on $x_1$ orbits that oppose the bar. So, we would expect that post-bar formation stars should have the opposite orientation. Since bright OH/IR stars are expected to be relatively young, potentially, an analysis of their kinematics and surface distribution could test this hypothesis.

## 3.4. THE STELLAR KINEMATICAL SIGNATURES OF A GALACTIC BAR

Do the kinematics of the Galactic bulge suggest that the Galactic bulge is barred? Present observations neither contradict nor confirm the existence of the bar. However, planned and on-going observations will provide an important test of the bar hypothesis and potentially provide important constraints on its dynamics and origin.

Different stellar populations will have different kinematic signatures in barred potentials depending upon whether they are primarily on box, short axis loop or long-axis loop orbits. If the Galaxy has a rapidly rotating bar with a pattern speed 81 km $s^{-1}$ kpc$^{-1}$ as suggested by Binney et al.(1991), then we would expect to see this figure rotation reflected in stellar motions. Attempts to find definitive evidence for such a rotation rate is confused by the fact that observed rotation rates may be due to a combination of figure rotation and streaming motion.

The stars that make up the backbone of the bar are primarily on box orbits. These orbits align along the major axis of the bar and are needed to support the potential. In a rotating bar, these orbits do have a sense of rotation. Thus, observations of a stellar population that is primarily composed of this orbital family should show a rotation rate similar to the rotation rate of the bar. It is intriguing that observations of older evolved populations show evidence for rotation rates consistent with the Outer Lindblad Resonance of the bar lying at $\sim 3.5$ kpc, the location suggested by the Binney et al. (1991) study of the stellar kinematics. Catchpole (1990) find a rotation rate of $14.2 \pm 5.2$ km $s^{-1}$ per degree for a sample of 52 Miras. Menzies (1990) find a rotation rate of $9.8 \pm 1.9$ km $s^{-1}$ per degree for the IRAS Miras in the bulge. Unfortunately, we do not yet have any self-consistent stellar models for the Galactic bar, so that it is hard to make detailed comparisons between observed kinematics and dynamical properties of the bulge.

Stars that form from the gas is the plane of the bar would have a completely different kinematic signature than stars that formed before or during bar formation. It is intruiging that the OH/IR stars, particularly the young OH/IR stars show much more rapid rotation, $(1 \text{ km } s^{-1} \text{ pc}^{-1};$ Habing *et al.* 1983, Winnberg *et al.* 1985), than other stellar populations. Perhaps this reflects that these stars are primarily on $x_2$ orbits, much like the gas, and have significant streaming motions. Since OH/IR stars are believed to be tracers of an intermediate age population, it is perhaps not suprising that they have very different kinematic properties from older stars.

As noted in Section 2, while minor axis rotation cannot occur in an axisymmetric galaxy, the presence of long-axis loop orbits makes it possible in a barred galaxy. Thus, the detection of minor axis rotation would be definitive evidence for triaxiality. N. Tyson, M. Rich and I have planned an observing program geared to detecting minor axis rotation in the bulge. By observing K giants along the minor axis, both above and below the plane, we hope to constrain the LSR motion and determine whether there is any evidence for minor axis rotation.

Just as particle physicists probe the structure of atomic nuclei by scattering particles off of targets, astronomers can probe the structure of the bulge by observing stars in the local neighborhood that have been scattered off of the Galactic bar. Low angular momentum stars in the solar neighborhood should be significantly affected by the presence of a stellar bar in the Galactic center. These stars will have plunged through the bar many times as they move on their orbits through the Galaxy. The bar will have a different orientation during every passage. Thus, the star will be randomly scattered and will not be able to stay on a regular orbit. In fact, we suspect that the star will eventually be scattered off of the orbit to another orbit that does not spend as much time in the solar neighborhood. K. Long and D. Spergel have been exploring the scattering of solar neighborhood stars off of the Galactic bar. Our preliminary investigations suggest that low angular momentum stars should be depleted in the solar neighborhood. Such a depletion would confirm the presence of the bar and allow an alternative way to measure the local circular speed. This scattering mechanism will also occasionally lead to the ejection of bulge stars into the local neighborhood. Perhaps, the Bulge supermetal rich stars found in the local neighborhood by Barbuy and Grenon (1990) suffered such a fate. Carlberg and Innanen (1987) have suggested that a similar mechanism could operate in an axisymmetric potential with a sufficiently large central density cusp. Recent photometric measurements of the Bulge (Kent *et al.* 1991), however, suggest that the central density distribution would not produce the large cusp assumed by Carlberg and Innanen.

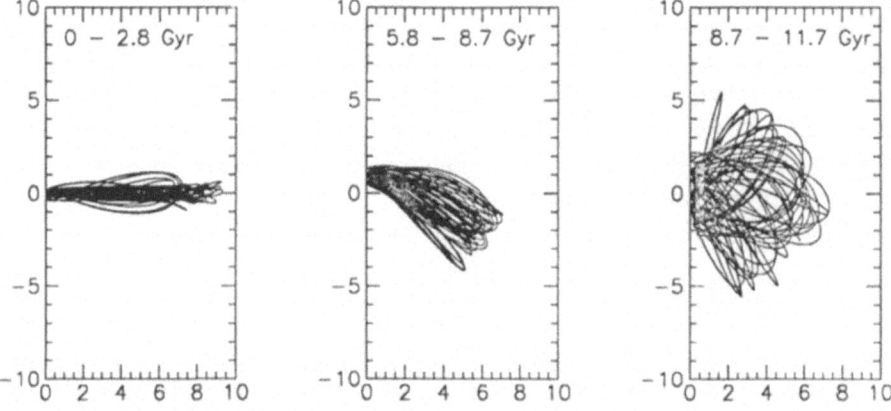

Fig. 6. The orbit of a star with initially low angular momentum in a barred galaxy.
The star's orbit is chaotic. The first panel shows the star's orbit for the first 4 Gyr.
The next panel shows the next 2 Gyr and the third panel shows the final 4 Gyr.
The star's orbit is scattered by the bar. This mechanism should deplete the number
of low angular momentum stars in the local neighborhood.

### 3.5. IS THE BAR TILTED?

Both photometric and kinematic evidence hint that the Galaxy bar is tilted
with it major axis pointing out of the plane of the Galactic disk.

In the Matsamuto et al. infrared maps, there is an asymmetric excess in
emission at positive latitude for $l > 0$ at at negative latitude for $l < 0$. This
asymmetry is apparent in figure 4 as a small region of negative excess. In
Blitz and Spergel (1991 b), we suggest that this excess is due to a tilt in
the Galactic bar and found direct photometric evidence for the tilt. There
is clear excess emission at both positive longitude and negative latitude, as
well as negative longitude and positive latitude. The excesses are clearly
coherent over a large area and are not due to random fluctuations.

There is already convincing evidence that the gas disk in the inner kiloparsec is tilted (Liszt and Burton 1980 and Sinha 1979). Perhaps, the simplest
evidence for a tilted elliptical gas distribution comes from the properties of
gas at forbidden velocities in the longitude-velocity diagrams. In an axisymmetric galaxy, in which gas moves on circular orbits, there should be no gas
in two of the quadrants in the $l - v$ plane ($l > 0$ and $v < 0$ or $l < 0$ and
$v > 0$), the "forbidden velocities". In our own Galaxy, there is a substan-

tial amount of gas in both forbidden velocity regions, a much more than expected from the velocity dispersion of the gas alone. The distribution in the forbidden velocity regions is consistent with expectations for the barred Galaxy model of Binney et al. (1991). Unlike the situation for most of the $l - v$ diagram, there is no distance ambiguity for the forbidden velocity gas: gas at $l > 0$ and $v < 0$ lies between us and the Galactic center, while gas at $l < 0$ and $v > 0$ lies on the far side of the Galactic center. Figure 7 shows that most of the gas on the near side is above the plane, while most of the gas on the far side of the Galaxy is below the plane. This systematic shift is seen over the entire range of longitudes for which there is a substantial amount of forbidden velocity gas $355° < l < 5°$ (see Burton and Liszt 1983). This systematic trend is apparent when the HI distribution is integrated over forbidden velocities (see figure 11 in Liszt and Burton 1978) and in the distribution of CO in the inner Galaxy (see figure 2 in Bally et al. 1989).

The near infrared continuum data will soon be available from the COBE satellite. This multicolor data will be an important test of the inferences made from the 2.4 $\mu$m map and will allow quantitative estimates of the scale length and axial ratio of all three axis of the Bulge.

The Galaxy's tilted Bulge may be typical of spiral galaxies, rather than an anomaly: careful analysis of the bulge of M31 by Ciardullo et al. (1988) suggest that the bulge of M31 is tilted with respect to the plane of its galactic disk. Why are bulges tilted? Pitesky (1991) has analyzed the interaction between an oblate halo, a bulge and a truncated exponential disk in an extension of the Sparke and Casertano (1988) model of warps as discrete modes. She finds that the bulge major axis tends to lie between the plane of the inner disk and the equatorial plane of the halo. Thus, a tilted bulge may be the natural complement of a warped disk. In order to simplify normal mode analysis, Pitesky assumes a rigid halo— it will be important to determine if the tilted bulge in her simulation would persist if the halo were dynamic. Ostriker and Binney (1989) suggest that the tilted gas distribution in the central regions of the Galaxy is due to the accretion of gas whose underlying angular momentum vector is not aligned with the underlying Galaxy. This theory would not explain the tilting of the stellar bulge, unless there is sufficent mass in the gas to torque the Galactic Bulge.

## 3.6. Are Bars Thick or Thin?

Numerical simulations of bars (Combes et al. 1990. Raha et al. 1990) suggest that strong bars are unstable to a "firehose" instability whose final product is a galaxy that appears bar-like from above, but peanut-shaped when viewed edge-on. Based on these simulations, Combes et al. (1990) propose that bars are not thin as suggested by Kormendy (1982), but rather are peanut-shaped. It is intriguing to think that these numerical simulations

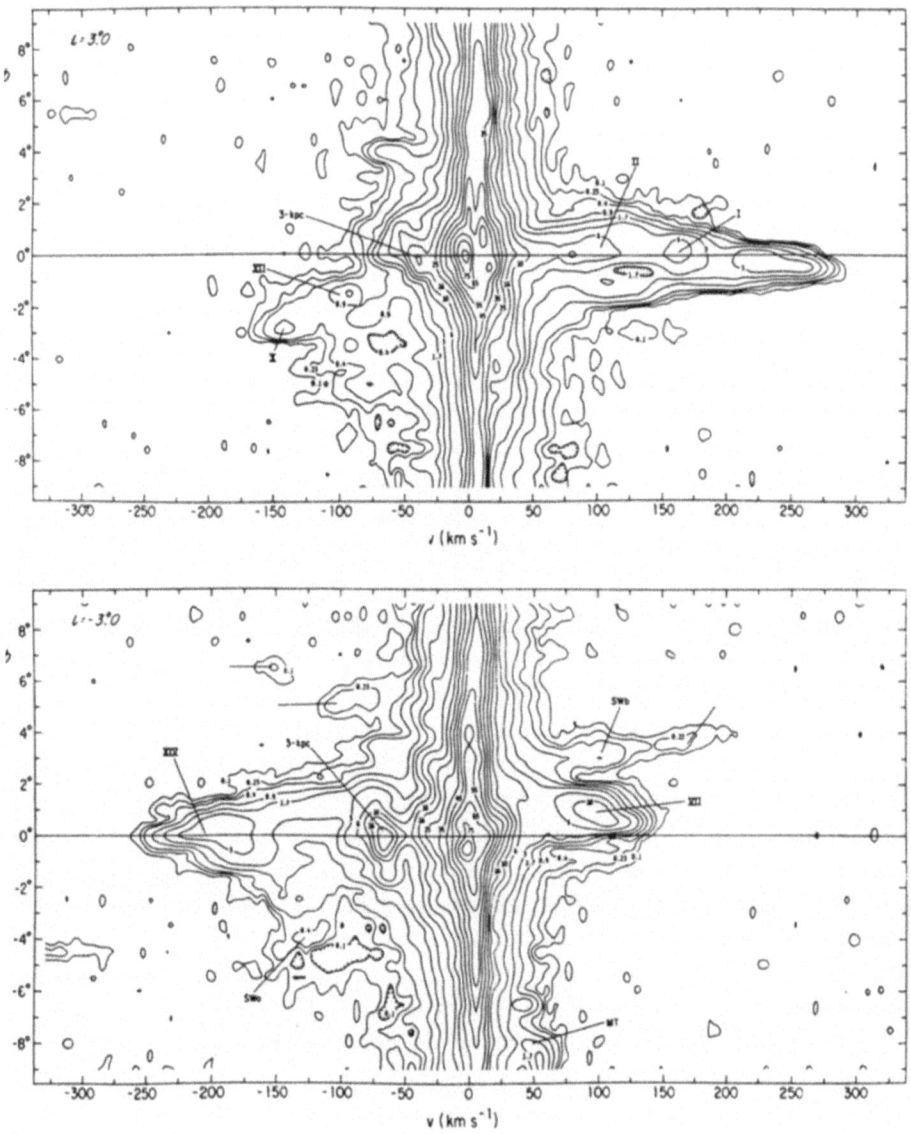

Fig. 7. Two slices in the latitude-velocity space corresponding to $l = 357°$ and $l = 3°$. Note that most of the gas at forbidden velocities ($v < 0$ for $l > 0$) lies below the Galactic plane, while most of the gas at forbidden velocities on the far side of the Galaxy ($v > 0$ and $l < 0$) lies above the Galactic plane. (From Burton and Liszt 1983)

may be describing the formation history of the Galactic Bulge. This formation picture would allow the bulge to form dissipatively as a disk and then be thickened by dynamical instability.

One of the classic arguments for thin bars is that a large fraction of all face-on galaxies have bars, which are not apparent in edge-on systems. Combes *et al.* (1990) argues that peanut-shaped galaxies may be barred systems viewed edge-on. Our Galactic Bulge may be such a system— a triaxial peanut-shaped bulge. It will be exciting to be able to compare the detailed kinematical predictions of the simulations with observations of the stars in the bulge. This will enable us to determine if the processes simulated in the computer reflect the physics of bar formation.

Just as HI observations of our own Galaxy can be used to probe its three-dimensional stucture, HI observations from external galaxies can also be powerful probes of its structure. HI observations of peanut-shaped galaxies can in principle determine whether they are triaxial or axisymmetric systems. While the terminal velocity curve cannot differentiate between these two possibilities, they imply different gas distributions in the $l - v$ plane of these edge-on systems.

### 3.7. THE FATE OF THE GAS AND THE SECULAR EVOLUTION OF THE BAR

Estimates of the mass of gas in the central 400 parsecs of the Galaxy vary widely largely because the surface density is dominated by the molecular gas, and various molecular mass tracers give conflicting estimates. The molecular gas in the inner regions can be traced by either the diffuse CO, $^{13}$CO, or CS emission which has been extensively mapped (Bally *et al.* 1987). The gas close to the center is manifested by strongly non-circular motions and is thus easy to distinguish from the disk gas. These highly non-circular velocities are observed only within about $3°$ of the center, implying that the most of the gas is confined to a radius of 400 pc from the center. If one assumes that the emissivity of CO is proportional to $N(H_2)$, then the surface density of the molecular gas is about 60 times higher than it is locally (Sanders *et al.* 1984). Mass estimates of the molecular gas alone suggest central masses of $2 \times 10^8 M_\odot$ about 10 – 20% of the total mass of molecular gas in the Milky Way. Evidently this large mass is concentrated within only 0.02% of the surface area of the disk.

A much lower mass estimate comes from the absence of diffuse high energy gamma-rays from the inner few degrees of the Galactic center (Blitz *et al.* 1985). The gamma-rays are produced from $\pi^0$ decay that results from the interaction of cosmic ray nucleons with the interstellar gas. In spite of the large surface density of CO and CS at the center, no gamma-ray emission is observed from the center in excess of the background coming from the disk. Either the cosmic ray density is low in the center, or the abundance of the trace species is much higher in the center than in the disk. It is known from the synchrotron emission that the density of cosmic ray electrons in the center is, if anything enhanced. Thus if the gamma-ray deficit is due to molecular abundance variations, the $3\sigma$ upper limit to the mass of molecular gas in the center is about $5 \times 10^7$ $M_\odot$. Circumstantial evidence, however, now implicates the cosmic rays as the source of the deficit. VLA observations of the Galactic center show that it is suffused with structures that imply poloidal magnetic fields. These fields may channel the cosmic rays out of the Galaxy, and inspite of their abundance, may have few interactions with the abundant molecular gas. Furthermore, the discovery of large scale diffuse CS emission, which requires relatively high densities to excite (compared to CO) is difficult to explain if it results merely from an abundance effect. Although the final story is not yet in, the best interpretation of all of the observations suggests that the high masses and surface densities implied by the CO observations are at least approximately correct.

Where did this gas come from? It seems unlikely that the gas was accreted from outside the Galaxy through the recent accretion of an LMC size companion. The presence of an old thin disk suggests that the Galaxy has not accreted a significant satellite recently (Toth and Ostriker 1992). If the Galaxy accreted a metal weak satellite with young stars in the last few Gyrs, then we would expect to detect Carbon stars in the Bulge. The absence of such stars and the metal abundance distribution observed in the K giants suggest a relatively merger-free past (Rich 1989).

There are two plausible sources for the gas in the Bulge: gas in the disk could be driven inwards by the bar or the gas may represent mass-loss from bulge stars. Bulge stars should be injecting roughly 0.1 $M_\odot$ yr$^{-1}$ into the interstellar medium (Frogel 1988). Faber and Gallagher (1976) and Bregman (1978) suggest that this gas will be driven out of bulges by Type Ia supernovae. These authors assume that almost all of the supernova energy will be converted into heating the Galactic Bulge gas. However, studies of the dynamics of supernova ejecta suggest that only $\sim 3\%$ of the kinetic energy of the ejecta is transfered into the interstellar medium (Spitzer 1978). Thus, clearing the Bulge of mass injected by stellar mass loss would require roughly 1 supernova every 7 years! Alternatively, the recent detection of X-ray line emission (Yamauchi et al. 1991) from the Bulge may imply that the gas in the Bulge is under extremely high pressure (Spergel and Blitz 1992). This higher pressure implies a higher efficency of coupling of supernova energy to the interstellar medium of the Bulge, so that a more traditional supernova rate ($\sim$ 1 supernova every 200 years in the Bulge) might be compatible with the clearing the Bulge of gas.

Where is it going? Binney et al. (1991) and Stark et al. (1991) have argued that this gas will spiral into the Galactic center within the next 1 Gyr. They note that dynamical friction will drive the observed molecular clouds inward.

This infall may either fuel the black hole posited in the Galactic center (see, *e.g.* Phinney 1989) and reactivate it or may spark another episode of star formation. If this gas is turned into stars in the next billion years, then the implied star formation rate, $0.02 M_\odot$ yr$^{-1}$ seems to be consistent with positing that the Bulge may undergo periodic starbursts (Rich 1991a,b). Rich (1991a,b) has argued that the Bulge contains intermediate age stars and also notes evidence for an intermediate age population in M31. This suggests that gas in the past may have been converted into bulge stars. Stars might be continually forming from the gas in the inner disk of the Galaxy and could be heated by resonant scattering (Pfenniger and Norman 1990) and eventually be detected at higher latitude through one of the high extinction windows.

Why isn't there gas in the bulge of M31? Observations of M31 find a 8 kiloparsec radius hole in its atomic and molecular gas distribution (see, e.g. Braun 1990). Somehow, M31 has eliminated all of its gas. Why do two similar galaxies have such different gas properties?

## 4. Galactic Spheroid

In this section, we discuss the dynamics of the Galactic spheroid, the metal-weak stellar population traced by the RR Lyrae stars, and the dynamics of the dark halo.

The Galactic spheroid appears to be a distinct component of our Galaxy, spheroid stars appear to differ from bulge stars in their kinematics, abundance distributions and in their spatial distribution. While the Bulge contains many metal-rich stars (Rich 1990, Tyson 1991) that are rapidly rotating (Harding 1990, Catchpole 1990, Menzies 1990), the halo stars in the local neighborhood are metal-poor and slowly rotating (Freeman 1987). Kent *et al.* (1991) suggest that photometrically the Galactic spheroid and bulge should be viewed as distinct stellar populations. In this section, we will discuss how observations of gas motions can probe the shape of the Galactic spheroid and the properties of the mysterious dark halo.

### 4.1. GAS KINEMATICS

In Blitz and Spergel (1991a), we proposed that the addition of a slowly rotating quadrupole term to the Galactic potential could account for the kinematics of the gas outside 4 kpc. We concluded that the gas between 4 kpc and the solar circle moves on concentric ellipses with nearly constant ellipticity, while the gas in the outer Galaxy moves on nearly circular orbits. This model is consistent with a large body of observational data, some of which had previously been a puzzle, including the near zero radial velocity of the 21-cm absorption feature toward the Galactic center, the $\sim$ 15 km/s offsets of the centroids of the molecular emission in the inner core of the Galaxy, the streaming motions of the spheroid stars and the lower values of the LSR corrections obtained from relatively local and stellar observations. Observations of the gas cannot reveal the source of the quadrupole, it may be either the stellar spheroid or the inner portion of the dark halo. In Blitz and Spergel (1991a), we emphasized the stellar spheroid as the source of the quadrupole and suggested that it was triaxial.

Observations of spheroid stars suggest that the spheroid is somewhat flattened. While observations of gas motions can be used to estimate the axial ratio of the intermediate axis to long axis $(b/a)$, stars can probe the flattening of the spheroid, $(c/a)$. Various authors reach somewhat different conclusions about the spheroid's shape. Using star counts of faint subdwarfs, Bahcall and Soniera (1984) find that $c/a = 0.80^{+0.20}_{-0.05}$. Sommer-Larson (1987) found $c/a = 0.8 \pm 0.15$ for the blue horizontal branch stars. Wyse and Gilmore (1988) using star counts towards the north Galactic pole conclude that $c/a < 0.5$. Estimates based on the kinematics of the spheroid stars are more uncertain. Binney and May (1986) suggest that $c/a \approx 0.25$ based on the highly anisotropic velocity dispersion tensor of the high velocity metal-poor stars. Vedel and Sommer-Larson (1990) find that $c/a \approx 0.7$ from their dynamical models of observations of blue horizontal branch stars.

One of the most striking pieces of evidence for the Galaxy not being axisymmetric comes from the observations of gas near the anti-center. If the gas in the outer Galaxy moved on circular orbits and the LSR did not have any radial component in its motion, then the gas at the Galactic anti-center should appear to be at rest. The observations, however, show that the gas in the anticenter is moving towards us at $\sim$ 15 km/s (Blitz and Spergel 1991a). The overall distribution of gas in the outer Galaxy shows a striking $m = 2$ distortion in the shape of the gas contours (see Figure 8). If the LSR were moving on a non-circular orbit, then this is just what we would expect to observe. In Blitz and Spergel (1991a), we argued that this non-circular motion could not be due to spiral arms, but rather was due to a large scale triaxial component, either a triaxial halo or a triaxial spheroid.

Alternatively, the outer Galaxy could be lopsided, which would also produce the asymmetry seen in the outer gas contours (Blitz and Spergel 1991a). Kuijken (1990) has explored the dynamics of an $m = 1$ distortion (a lopsided galaxy) and has argued that it is a preferable solution to the asymmetries in the gas distribution. In order to explain the $l - v$ diagram, this asymmetry must be the distribution of mass in the Galactic halo. Dynamically stable models of such a lopsided galaxy have not been constructed. However, numerous galaxies are observed to have an irregular distribution of gas in their outer regions (Baldwin, Lynden-Bell & Sancisi 1980). We do not know whether these $m = 1$ asymmetries in these external galaxies are due to asymmetries in the distribution of dark matter or are due to the gas

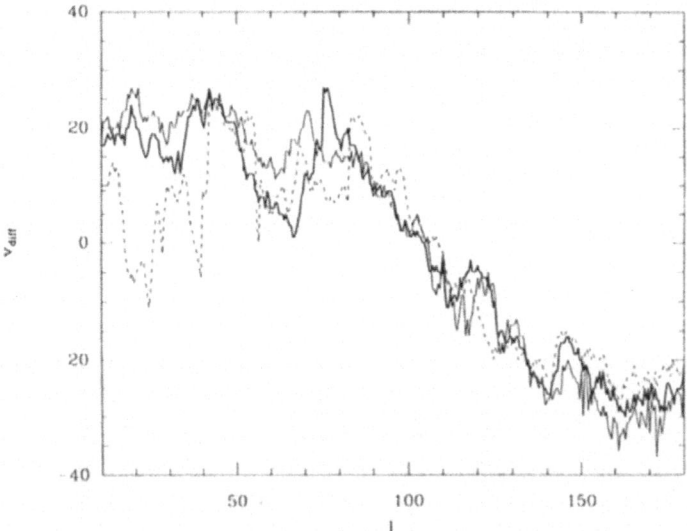

Fig. 8. The difference between the contours in the $0° < l < 180°$ and $180° < l < 360°$ regions, $v_{diff}$. The light solid line corresponds to the 1K isotherm in Figure 1. The heavy solid line corresponds to the 2K contour. The dashed line corresponds to the 4K contour (From Blitz and Spergel 1991a).

having not yet settled into an equilibrium.

The proposed observations of stars along the minor axis of the Galactic Bulge, outlined in the previous section, could provide the definitive test of these two models. If the asymmetries in the outer gas contours is due to LSR motion, then we should be moving away from the stars along the minor axis. If on the other hand, the LSR is moving on a circular orbit and the gas in the outer Galaxy is moving on a lopsided orbit, then there should be no motion with respect to the LSR.

Can we use the gas distribution to discriminate between an $m = 1$ mode and an $m = 2$ mode as the source of the irregularities seen the gas distribution? The closed orbits will cause a different spatial distribution of the HI in each case. An $m = 1$ mode will have more gas in the third and fourth Galactic quadrants than in the first and second (these quadrants are specified by the negative velocities of the gas seen in the Galactic anticenter). An $m = 2$ mode will cause gas to pile up in the second and fourth Galactic quadrants. Thus, it is possible to distinguish between these two cases by looking at the differential column densites of HI measured with respect to the Sun-center line. If one subtracts the column density at a longitude in the third and fourth Galactic quadrant from an equivalent co-longitude in the first and second, one would expect an $m = 1$ to produce an even function, and an $m = 2$ to produce an odd function as a result of the differences in gas distribution.

Figure 9 is such a plot, where N(diff)% is the column density difference (expressed as a percentage of the N(HI) at longitudes $> 180°$) between the two Galactic hemispheres plotted as a function of longitude. The figure clearly shows that the difference function is in fact odd. However, it is important to be sure that local features do not contaminate the results. To do this, we simply take the differential column densities starting at larger and larger distance from the solar circle. This is equivalent to integrating the HI emission from a given non-zero velocity to infinity, and again plotting N(diff)%; larger values of the velocity correspond to larger distances from the solar circle. The figure shows that the character of the plot does not change even for values of the velocity as large as 30 km $s^{-1}$ , and the oddness of the plot is determined largely by gas at large distances from the Sun.

The scale height of the HI layer as a function of azimuthal position and velocity may provide another test to discriminate between an $m = 1$ and an $m = 2$ mode. Merrifield (1992) demonstated how the scale height of the HI layer can be used as method for determining the rotation curve of the Milky Way outside the solar circle. Merrifield assumes that the scale height of the gas layer is independent of azimuth and then solves for the layer thickness and distance as a function of velocity. This analysis shows systematic differences between positive and negative longitudes which are perhaps due to the outer Galaxy being either lopsided or triaxial.

## 4.2. PHOTOMETRIC SIGNATURE

Does the distribution of stars show any evidence for non-axisymmetry outside of the inner 2 kiloparsecs? Weinberg (1991) has developed a new algorithm for determining the lowest order harmonics of the stellar density distribution about the Galactic center and applied it to a sample of AGB stars inside the solar circle. His sample consists of variable IRAS sources with 12 $\mu$m fluxes greater than 2 Jy. He restricts his sample to stars whose colors are indicative of dusty AGB stars. Weinberg finds that there is an excess in star counts and suggests that the AGB stars trace a bar whose semi-major axis is $\sim 5$ kpc and has a position angle of $-36°$.

Are Weinberg's (1991) results consistent with the stellar bar discussed in Section 3 and the gas kinematics discussed in Section 4? Probably not.

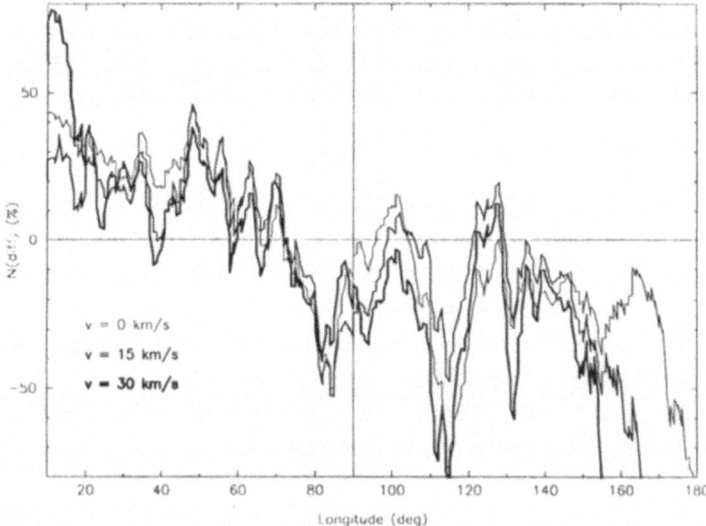

Fig. 9. The difference between the column density of HI at $180° < l < 360°$ and $0° < l < 180°$ as a fraction of N(HI) in the range $180° < l < 360°$. The three different curves plot three different velocities from which N(HI) is integrated to infinity. As the velocity increases, we omit gas at increasingly large distance from the solar circle. It is clear that in all cases, the function N(diff)% is an odd function of longitude implying that the HI in the Milky Way is distributed in an $m = 2$ mode.

If there is a rapidly rotating inner bar, then there would not be any stable radial orbits near the outer Linblad resonance. Thus, Weinberg's model requires that corotation is beyond 5 kiloparsecs, which is not compatible with either the gas in the inner Galaxy or the shape of the terminal velocity curve. The radial extent of the quadrupole component found by Weinberg is very different from the infrared radial profile of the Galactic Bulge (Kent *et al.* 1991).

I suspect that the Weinberg's detection of an asymmetry in the AGB distribution is a discovery of an elliptical thick disk, rather than the detection of a large-scale stellar bar. The stellar response to a quadrupole potential depends upon the orbital family that is dominant in a given stellar component. If the stars are mostly on box orbits, then the major axis of their density distribution should align with the potential. If stars are mostly on loop orbits, then the major axis of their density distribution should be perpendicular to the major axis of the potential. Stars on box orbits should show asymmetric drifts of roughly the circular speed, $\sim 225$ km/s. Stars that are primarily on loop orbits should have much smaller asymmetric drifts. te Lintel-Hekkert and de Jonghe (1992) have analyzed the orbits of the OH/IR stars in their IRAS selected sample. They concluded that these stars are part of a thick disk population and have an asymmetric drift of only 20 km/s. If we assume that the AGB stars are part of a thick disk, then its orientation is consistent with the orientation of the gaseous disk suggested in Blitz and Spergel (1991a).

Kent *et al.* (1991) photometry of the Galaxy identifies a thick disk component whose contribution to the light peaks at 5 kpc. It is tempting to identify this component with the AGB sample of Weinberg and the OH/IR sample of te Lintel-Hekkert *et al.* (1991). If the thick disk is indeed elliptical, then it should appear dimmer at negative longitudes. We eagerly await the analysis of the COBE Galaxy maps, which extend far deeper than is visible in the Hauser *et al.* (1990) photograph, and will provide important clues to the shape of both the thick disk and the spheroid.

### 4.3. STELLAR KINEMATICS

Any large-scale asymmetry in the Galaxy should be reflected in the kinematics of stars in the local neighborhood. In an axisymmetric Galaxy, the vertex of the velocity ellipsoid should point towards the Galactic center and the velocity ellipsoid should show no dependence on azimuth (*i.e.*, Oort's C constant should be zero). An $m = 2$ distortion, triaxiality, should produce both a vertex deviation and a non-zero value for the Oort C constant. Recently, Kuijken and Tremaine (1991) have compiled and analyzed observations of stellar motions in the solar neighborhood. They concluded that the stellar data do not show strong evidence for large-scale asymmetries, however, the data are not yet good enough to rule out the $m = 2$ distortions suggested by Blitz and Spergel (1991a).

Observations of stars in the outer Galaxy are another tracer of the Galactic potential. If the Galaxy has a rotating stellar halo with an ILR at $\sim 2R_0$, then closed orbits should become increasingly elliptical as they approach the Inner Linblad Resonance (ILR). As they cross the ILR, the orientation of the ellipse changes by 90° implying that the non-circular component of the stellar motion changes sign. Kuijken (1991) argues that this increase in outward motion with radius towards the ILR and the subsequent sign reversal should be apparent in the gas. Gas streamlines cannot cross, thus, rapid variations in gas radial velocity will be damped by dissipation— a result confirmed by hydrodynamical simulations by Bies (1989). On the other hand, stellar systems can move on intersecting orbits since stars are not collisional. The K giant survey of Lewis and Freeman (1989) may show a hint of this effect. They find that the K giants near the anticenter appear to deviate systematically from the LSR motion: at $R = 1.5R_\odot$, $1.75R_\odot$ and $2.1R_\odot$, they find that the K giants are moving away from the LSR at $13 \pm 4.2$ km/s, $16 \pm 4.6$ km/s and $5.7 \pm 4.5$ km/s implying either an inward motion of the LSR or perhaps, an outward motion increasing with distance from the galactic center. The test of this hypothesis would be observations of K giants at $\sim 2.5 R_\odot$: if the rotating stellar spheroid model is correct, then these stars should be moving *towards* the LSR at $\sim 15$ km/s.

Another potentially powerful probe of the Galactic potential is the carbon star survey of the late Marc Aaronson and his collaborators (Aaronson *et al.* 1989, Aaronson *et al.* 1990). When complete, these data will be helpful in constructing a coherent model of the Galaxy.

## 5. Conclusion

There is mounting evidence that our Galaxy is not axisymmetric, but rather a complex triaxial system. The Milky Way appears to have a central triaxial component that may be rapidly rotating. There is also evidence that either the Galactic spheroid or inner halo is triaxial or the Galaxy is lopsided. These observations of the three dimensional structure of the Milky Way are of more than cartographic interest, for they provide hints about the Galaxy's formation and evolution. (It is not clear whether we should divide the Galaxy's history into these distinct phases since Galaxy formation may be an ongoing process.)

The shape of Galactic halo is a sensitive probe of the initial conditions that led to galaxy formation. Carlberg and Dubinski (1991) found in their simulations of galaxy formation in the CDM scenario that most galaxy halos were highly triaxial. I suspect that this will be true in any heirarhical scenario in which the halo forms dissipationlessly from Gaussian initial conditions. Blitz and Spergel (1991a) and Kuijken (1991) have shown how HI observations can be used to probe not only the monopole term in the potential, but also higher order terms. Observations of the Galaxy's triaxiality combined with observations of the gas motions in the outer regions of other galaxies are a potentially powerful test of galaxy formation scenarios.

If the Bulge is a rapidly rotating triaxial system, then it should be viewed as a physically distinct system from a slowly rotating triaxial galactic spheroid. These two systems might have remarkably different formation histories. Binney (1991) notes that the triaxial spheroid suggested in Blitz and Spergel (1991a) is akin to an elliptical galaxy. A triaxial system would form naturally in hierarchical models where the spheroid formed through the merger of satellite galaxies (Frenk and White 1991). The bulge/bar, on the other hand, might have grown through the bar and firehose instabilities our of a disk that extended into the center of the Galaxy. Thus, dissipation could have played an important role in the formation of the Bulge, which would account for its high phase space density. It is less clear how dissipational galaxy formation would lead to an elliptical thick disk that is aligned perpendicular to the major axis of the spheroid.

Evidence for triaxiality in the Galactic Bulge should stimulate astronomers to rethink the classic notion that the Bulge has a static memory of an early dissipationless collapse. Recent work by Hernquist and Weinberg (1992) found that a bar can transfer its angular momentum to a spheroid or halo in a few dynamical times. This suggests that a stellar bar is a dynamically evolving object. Thus, either bars are short-lived phenomena or the bar is constantly replacing angular momentum lost to the halo with angular momentum taken from the disk. If disk gas is constantly flowing into the Bulge, the bulge star formation should be an on-going processe. The molecular gas disk observed in the inner 500 parsecs of our Galaxy may not be a transient phenomenon, but rather is long-lived. Stars would be constantly forming in this thin gas layer. These stars would not be apparent to optical observers until dynamical heating processes increase their dispersion so that they can be observed in Baade's window and other Bulge windows. These stars may, however, be potentially detectable in the infrared. A signature of this would be the detection of intermediate age stars in the Galactic center region.

Observations of the kinematics of stars in the Galactic bar will help resolve many of the questions about the dynamics of bars. The Galactic bar is clearly a "thick bar", not thin. By determining which stellar orbits make up the bar, we should be able to test various hypothesis about bar formation and evolution.

If the triaxiality of the bulge and spheroid are driving spiral structure, then this may have important implications for the chemical evolution of the Galactic disk. Gas crossing spiral shocks loses angular momentum. Given the velocity changes observed across the spiral arms of our Galaxy, gas should be driven inwards on roughly 20 dynamical times, a timescale short compared to the age of the Galaxy. This constant infall means that the chemical evolution of various zones of the Galaxy are linked and that the Galactic disk should not be viewed as a closed box. This slow infall of gas will not produce significant X-ray flux— one of the classic objections to accretion models (*e.g.* Binney and Tremaine 1987).

Detailed observations and analysis of the complex structure of our Galaxy affords us the opportunity to learn the physics of galaxy formation and evolution and to unravel our own history.

## Acknowledgements

I would like to thank Neil Tyson, Martin Weinberg and Mike Rich for helpful conversations and Neil Tyson for a careful reading of an earlier draft. I would particularly like to thank Leo Blitz for an enjoyable collaboration upon which this article is based and for his contributions to this article. My research is supported in part by an A.P. Sloan Foundation Fellowship, by NSF grants

AST 88-58145 (PYI) and AST 91-17388 and by NASA grant NAGW-2448.

## References

Aaronson, M., Blanco, V.M., Cook, K.H. and Schechter, P.L.: 1989. *Astrophys. J. Suppl.*, **70**, 637.

Aaronson, M., Blanco, V.M., Cook, K.H., Olszewski, E.W. and Schechter, P.L: 1990. *Astrophys. J. Suppl.*, **73**, 841.

Bahcall, J.N. and Soniera, R.M.: 1984. *Astrophys. J. Suppl.*, **55**, 67.

Baldwin, J.E., Lynden-Bell, D.E. and Sancisi, R.: 1980. *Mon. Not. R. astr. Soc.*, **193**, 313.

Bally, J., Stark, A.A., Wilson, R.W. and Henkel, C: 1987: *Astrophys. J. Suppl.*, **65**, 13.

Bally, J., Stark, A.A., Wilson, R.W. and Henkel, C: 1988: *Astrophys. J.*, **324**, 223.

Barbuy, B. and Grenon, M. 1990. in *ESO/CTIO Workshop on Bulges of Galaxies*, Eds. Jarvis, B.J. and Terndrup, D.M. (Munich:ESO).

Bies, W: 1989, Princeton University senior thesis.

Binney, J., Gerhard, O. E., Stark, A. A., Bally, J. & Uchida, K. I.: 1991. *Mon. Not. R. astr. Soc.*, **252**, 210.

Binney, J. & May, A.: 1986. *Mon. Not. R. astr. Soc.*, **218**, 743.

Binney, J. & Tremaine,S.: 1987. *Galactic Dynamics*, Princeton Univ. Press, NJ.

Blanco, V.M. and Terndrup, D.M.:1989. *Astronom. J.*, , **98**, 843.

Blitz, L., Bloemen, J.B.G.M., Hermsen, W. & Bania, T.M., Å, **143**, 267 (1985).

Blitz, L. & Spergel, D.N.: 1991a. *Astrophys. J.*, **370**, 205.

Blitz, L. & Spergel, D.N.: 1991b. *Astrophys. J.*, **379**, 631.

Braun, R.: 1990. *Astrophys. J. Suppl.*, **72**, 775.

Bregman J.N.: 1978. *Astrophys. J.*, **224**, 308.

Burton, W.B., & Liszt, H.S.: 1978, *Astrophys. J.*, , **225**, 815.

Burton, W.B., & Liszt, H.S.: 1983, *Astron. Astrophys. Suppl.*, **52**, 63.

Carlberg, R.A. & Dubinski, K.A.: 1991, *Astrophys. J.*, **378**, 496.

Carlberg, R.A. & Innanen, K.A.: 1987. *Astronom. J.*, **94**, 666.

Catchpole, R.M.:1990. in *ESO/CTIO Workshop on Bulges of Galaxies*, Eds. Jarvis, B.J. and Terndrup, D.M. (Munich:ESO).

Ciardullo, R., Rubin, V.C., Jacoby, G.H., Ford, H.C., & Ford, K.C.: 1988. *Astronom. J.*, **95**, 438.

Combes, F., Debbasch, F., Friedli, D. & Pfenniger, D.: 1990. *Astron. Astrophys.*, **233**, 82.

Contopoulos, G. & Grosbol, P.: 1989. *Astron. Astrophys. Rev.*, **1**, 261.

Contopoulos, G. & Mertzanides, C.: 1977. Å**61**, 477.

de Vaucouleurs, G: 1964, in *IAU Symposium 20: The Galaxy & the Magellanic Clouds*, Kerr, F.J. & Rodgers, A.W.,Ed.,(Sydney: Australian Academy of Science), p: 195.

Faber , S.M. & Gallagher, J.S.: 1976. *Astrophys. J.*, , **204**, 365.

Freeman, K.: 1987. *Ann. Rev. Astron. Astrophys.*, **25**, 603.

Frenk, C.S. and White, S.D.M. 1991: *Astrophys. J.*, **252**, 75.

Frogel, J.A.: 1988. *Ann. Rev. Astron. Astrophys.*, **26**, 53.

Gerhard, O. E. & Vietri, M.: 1986. *Mon. Not. R. astr. Soc.*, **223**, 377.

Gilmore, G., King, I. & Searle, I.: 1990. *The Milky Way as a Galaxy*, ed. by R. Buser & I. King (Univ. Science Books: Mill Valley, CA).

Habing, H.J., Olnon, F.M., Chester, T., Gillet, F., Rowan-Robinson, M. & Neugebauer, G.: 1983. *Astron. Astrophys.*, **152**, L1.

Harding, P.: 1990. in *ESO/CTIO Workshop on Bulges of Galaxies*, Eds. Jarvis, B.J. and Terndrup, D.M. (Munich:ESO).

Harmon, H. & Gilmore, G.: 1988. *Mon. Not. R. astr. Soc.*, **235**, 1025.

Hauser, M.G., *et al.* : 1990: NASA photograph G90-03046.

Hernquist, L. & Weinberg, M.: 1992, submitted to *Astrophys. J.*

Kent, S.M., Dame, T.M. & Fazio, G.: 1991. *Astrophys. J.*, **378**, 131.

Kuijken, K: 1991. in *Warped Disk & Inclined Rings around Galaxies*, Casertano, S., Sackett, P.D. and Briggs, F.H., Ed. (Cambridge Univ. Press.), p. 159.

Kuijken, K. & Tremaine, S.: 1991. CITA preprint, to appear in *Dynamics of Disk Galaxies*.

te Lintel Hekkert, P. Caswell, J.L.,& Habing, H.J.: 1991. *Astron. & Astrophys. Suppl.* **90**, 327.

te Lintel Hekkert, P., Dejonghe, H. & Habing, H.J.: 1992. submitted to Å

Lees, J. & Schwarzschild, M. 1991. Princeton University Preprint.

Lewis, J.R. & Freeman, K.C.: 1990. *Astronom. J.*, **97**, 139.

Liszt, H.S., & Burton, W.B.: 1980.*Astrophys. J.*, **236**, 779.

Matsumoto, T., Hayakawa, S., Koizumi, H., Murakami, H., Uyama, K., Yamagami, T. & Thomas, J. A.: 1982. In *The Galactic Centre*, Eds. G. R. Riegler & R. D. Bl&ford, p. 48.

Menzies, J.W.: 1990. in *ESO/CTIO Workshop on Bulges of Galaxies*, Eds. Jarvis, B.J. and Terndrup, D.M. (Munich:ESO).

Merrifield, M.R: 1992, CITA preprint.

Miralda-Escude & Schwarzschild, M.: 1989. *Astrophys. J.*, **339**, 752.

Mulder, W.A .& Liem, B.T.: 1986. *Astron. Astrophys.*, **157**, 148.

Nakada, Y., Deguchi, S., Hashimoto, O., Izumiura, H., Onaka, T., Sekiguchi, K. & Yamamura, I.: 1991. *Nature*, **353**,140.

Ostriker, E.C. & Binney, J.J: 1989. *Mon. Not. R. astr. Soc.*, **237**, 785.

Peters, W.L: 1975. *Astrophys. J.*, **195**, 617.

Phinney, E.S.: 1989. in *IAU Symposium No. 136: The Center of the Galaxy*, Morris, M., ed. (Dordrecht: Kluwer Acad. Publ.), p. 543.

Pitesky, J: 1991. in *Warped Disk & Inclined Rings around Galaxies*, Casertano, S., Sackett, P.D. and Briggs, F.H., Ed. (Cambridge Univ. Press.), p. 215.

Pfenniger, D. & Freideli, D: 1991. Å252, 75.

Pfenniger, D. & Norman, C: 1990. *Astrophys. J.*, **363**, 391.

Raha, N., Sellwood, J.A., James, R.A. & Kahn, F.D: 1991. *Nature*, **352**, 411.

Rich, R.M.: 1989. in IAU Symposium No. 136: The Center of the Galaxy, Morris, M., ed. (Dordrecht: Kluwer Acad. Publ.), p. 63.

Rich, R.M.: 1990. *Astrophys. J.*, **362**, 604.

Rich, R.M.: 1991a. in *IAU Symp. 149 Stellar Populations*, A. Renzini & B. Barbuy eds., in press.

Rich, R.M.: 1991b. in *the proceedings of the Taiwan Astrophysics Workshop*, A. Fillipenko, ed., in press.

Sanders, D.B., Solomon, P.M. & Scoville, N.Z. 1984, *Astrophys. J.*, **276**, 182,

Sanders, R.H. & Huntley, J.M.: 1976. *Astrophys. J.*, **209**, 53.

Sinha, R.P.: 1979. in *The Large Scale Characteristics of the Galaxy*, W.B. Burton, ed., p.341.

Sommer-Larson, J.: 1987. *Mon. Not. R. astr. Soc.*, **227**, 21P.

Sparke, L. & Casertano, S: 1988. *Mon. Not. R. astr. Soc.*, **234**, 873.

Spergel, D.N. & Blitz, L.: 1992. to appear in *Nature*.

Spitzer, L., "Physical Process in the ISM", (Wiley Scientific: New York), 1978.

Stark, A.A., Bally, J., Binney, J. & Gerhard, O. E.: 1991. *Mon. Not. R. astr. Soc.*, **247**, 678.

Toth, G. & Ostriker, J.P.: 1992. *Astrophys. J.*, **389**, 5.

Tyson, N.D. 1991: Ph.D. Thesis, Columbia.

van Albada, G.D: 1985, in *The Milky Way Galaxy*, ed. H. van Woerden, R.J. Allen & W.B. Burton (Reidel: Dordrecht), p547.

Vedel, & Sommer-Larson: 1990. *Astrophys. J.*, **350**, 104.

Weinberg, M.: 1991. U. Mass. preprint.

Winnberg, A., Baud, B., Matthews, H.E., Habing, H.J., & Olnon, F.M.: 1985. *Astrophys. J. Letters*, **291**, L45, (1985).

Wyse, R. and Gilmore, G.: 1988. *Astronom. J.*, **91**, 855.

Yamauchi, S. *et al.*: 1991 *Astrophys. J.*, **365**, 532.

Yuan, C.: 1984. *Astrophys. J.*, **281**, 600.

# THE SHAPE OF THE BULGE FROM IRAS MIRAS

PATRICIA WHITELOCK

*South African Astronomical Observatory,*
*P O Box 9, Observatory, Cape 7935, South Africa*

and

ROBIN CATCHPOLE

*Royal Greenwich Observatory,*
*Madingley Road, Cambridge CB3 OEZ, United Kingdom*

**Abstract.** The number distribution of Miras in the Bulge as a function of distance modulus indicates that the half of the Bulge which is at positive galactic longitude is closer to us than the other half. The observed stellar distribution can be modelled with a triaxial ellipsoidal distribution of stars which we refer to as a 'bar' which is tilted at roughly 45° to the line of sight. Furthermore this bar is rotating in a manner consistent with expectations from the kinematics of the galactic centre gas.

## 1. Introduction

Mira variables provide a rather sensitive tool for the study of galactic structure. Oxygen-rich Miras in particular provide a potentially very precise tracer of metal-rich populations. Observationally we recognize these stars as large-amplitude pulsating M stars with a typical visual magnitude range in excess of 2.5 mag and periods of 100 days up to around 2000 days. From the standpoint of stellar evolution they represent a rather short phase in the evolution of low-(and perhaps intermediate-) mass stars prior to the formation of a planetary nebula. Particularly important for detecting them over large distances is the fact that they are at the maximum bolometric luminosity that they will ever reach, $L \gtrsim 2.10^3 L_\odot$, the exact value depending on initial mass and metallicity (Feast & Whitelock 1987).

One of the easiest measurements we can make of a Mira is its pulsation period and there are various correlations which suggest that this is a good indicator of the stellar population to which it belongs. Within the globular clusters we find that in the few cases where a cluster has more than one Mira these are all of nearly the same period. The periods of Miras in clusters correlate very well with the cluster metallicity. The kinematics of the Miras in the solar neighbourhood are a function of period; stars with periods of 200 day belong to an intermediate population II or thick-disk population as do the metal-rich globular clusters. There is a gradual change in properties up to the 400 day objects which have characteristics typical of the old disk (Feast 1963). This indicates a relationship of the pulsation period with either age, mass and/or abundance. It is for this reason that they provide a precise

103

*L. Blitz (ed.), The Center, Bulge, and Disk of the Milky Way,* 103–110.
© 1992 *Kluwer Academic Publishers.*

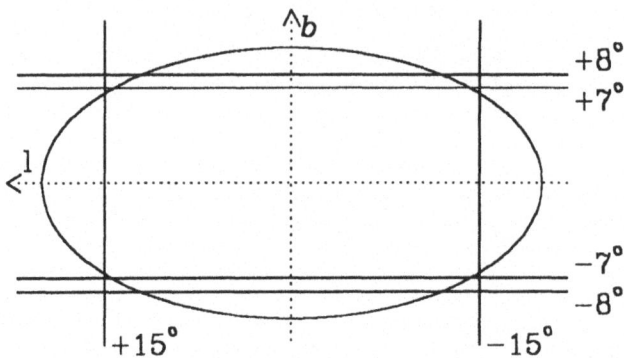

Fig. 1. The two strips across the Bulge surveyed for extreme Miras

tracer of a particular population. Examine Miras within a particular period range and you probe a particular mass/metallicity range.

Miras show a clear period-luminosity relation in both bolometric and K light. Miras in the LMC, SMC, galactic globular clusters and the Baade windows all obey the same PL relation. The available evidence suggests that OH/IR stars also obey this relation. Unfortunately there are few OH/IR stars with accurate distances, so it is not possible to be sure that this is an accurate representation for all such stars (see Whitelock, Feast & Catchpole 1991 and references therein).

## 2. IRAS Miras in the Bulge

Although bolometrically luminous, many Miras and OH/IR sources are optically faint or invisible because they are surrounded by thick dust shells as a consequence of their high mass-loss rates. Such objects are powerful emitters at 12 and 25 $\mu$m and were detected to considerable distances in the IRAS survey. Whitelock et al. (1991) recently published the results of a search for high-mass-loss Miras within the Galactic Bulge. They selected IRAS sources by colour from two strips across the Bulge, as illustrated in Fig. 1, and monitored them for several years in the near infrared (JHKL). The Miras can be distinguished with a high degree of success on the basis of their near-IR colours alone but monitoring for periodicity confirms or rejects the initial classification. The end result was 113 Miras, 22 M giants and 6 other objects. Of the 113 Miras periods were derived for 104, the other 9 being either too confused or in regions of very high interstellar extinction.

The absolute magnitudes of individual Miras for which periods have been determined can be derived from a period-luminosity relation, such as equation (2) of Whitelock et al. (1991). With the combination of near-infrared and IRAS data we have fluxes covering the spectral range, 1.25 to 25 $\mu$m, from which the apparent bolometric luminosity can be calculated by integrating under the flux curve. The distance to each individual Mira can therefore be estimated.

Whitelock *et al.* produced a plot (their Fig. 11) of the number of Miras per 0.2 mag bin as a function of distance modulus. They then modelled this number-density distribution using ellipsoidal distributions of stars with a variety of power laws and axial ratios. They compared the results with those found by Terndrup (1988) and Blanco & Terndrup (1989) for the distribution of late-M stars (the progenitors of Mira variables) across the Bulge. The net result was that the distribution across the Bulge was more compact than the distribution in depth. The simplest interpretation of this as pointed out by Feast & Whitelock (1990) is that the Bulge is bar shaped.

Harmon & Gilmore (1988) remarked on the longitudinal asymmetry in the distribution of IRAS sources in the Bulge and suggested that triaxiality of the Bulge might offer an explanation. Another notable point about the IRAS objects surveyed by Whitelock *et al.* was their asymmetry in Galactic longitude with respect to the Galactic Centre. Of the 141 objects obeying the initial selection criteria 85 were at positive galactic longitudes and only 56 were at negative longitudes. There was no clear north/south asymmetry; 71 were at positive galactic latitudes and 70 at negative latitudes. Early investigators were disinclined to offer detailed explanations for such asymmetries among IRAS sources because of suspicions that such peculiarities might be, at least in part, an artifact of the IRAS catalogue. Nakada *et al.* (1991) discussed the longitudinal asymmetry of the IRAS sources and interpreted it as the consequence of a bar-shaped Bulge.

It has been known for a long while that the non-axisymmetric gravitational potential of a triaxial bulge would provide a good explanation of the gas kinematics in the galactic centre region (e.g. de Vaucouleurs 1964). Gerhard & Vietri (1986) suggested that it was only possible to reconcile the galactic centre gas morphology, infrared observations of the Bulge (Matsumoto *et al.* 1982) and the local density of spheroid stars, if the Galactic Bulge were non-axisymmetric. More recently Blitz & Spergel (1991) have modelled the Matsumoto *et al.* infrared scans with a triaxial stellar bar and pointed out that the new COBE infrared data supported their conclusion.

## 3. Asymmetry in the Bulge

Figure 2a is a plot of the distance distribution of the IRAS Miras on either side of the centre (data from Whitelock *et al.* 1991). The solid triangles and open circles show the distributions for Miras with positive and negative longitudes, respectively. It is clear from this figure that the peak of the distribution at negative longitudes is further from us than the peak on the other side. Figure 2b shows the longitude distribution of the Miras. Any

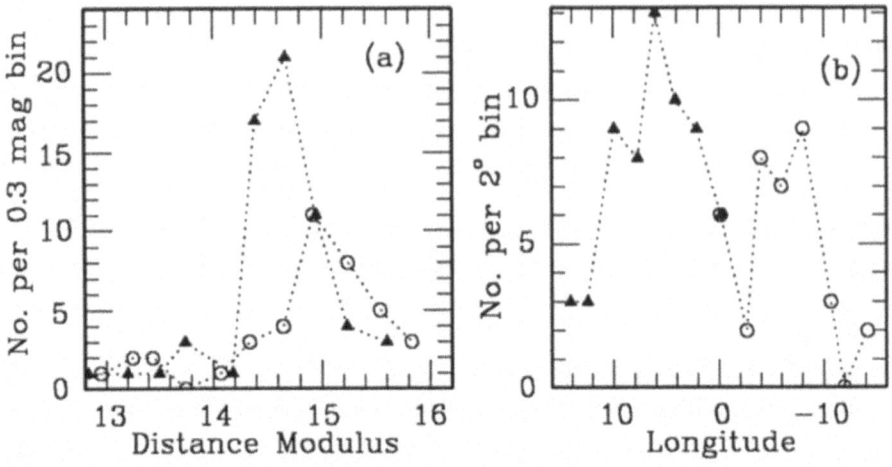

Fig. 2. (a) The distribution of IRAS Miras in distance modulus. The open circles and closed triangles represent those with negative and positive longitudes respectively. (b) The latitude distribution of the same sources.

model of the stellar distribution within the Bulge must fit this as well as the distance distribution of Fig. 2a. Of the 9 IRAS Miras for which periods, and therefore distances were not determined, 6 have negative and 3 have positive longitudes. It is very unlikely that there inclusion would significantly alter the picture.

If the distributions of IRAS Miras at positive and negative latitudes are compared then the peaks do occur at the same distance from the centre. Although as mentioned above the numbers of IRAS sources above and below the plane were almost identical, slightly more non-Miras were found above the plane. In the final analysis therefore the negative latitude group contains more Miras than the other one. The interstellar extinction is higher at positive than at negative galactic latitudes (see e.g. Lynds 1962 Fig. 2 and Whitelock *et al.* 1991) and this may contribute to the problem. A much larger sample of Miras is necessary before any firm conclusions can be drawn about these differences.

A relatively simple way of modelling the Bulge is with an ellipsoidal stellar distribution, which can be triaxial. In this case the number of stars per unit volume (N) can be expressed as:

$$N \propto exp - (x^2/x_o^2 + y^2/y_o^2 + z^2/z_o^2)^{1/2} \,, \tag{1}$$

where $x$, $y$ and $z$ are the distances from the center of the galaxy along the three axis of the ellipsoid and $x_o$, $y_o$ and $z_o$ are the scale lengths in each direction. It is assumed that the long axis ($x$) of this ellipsoid is tilted at some angle, $\theta$, to the line of sight and lies in the plane of the sky for which

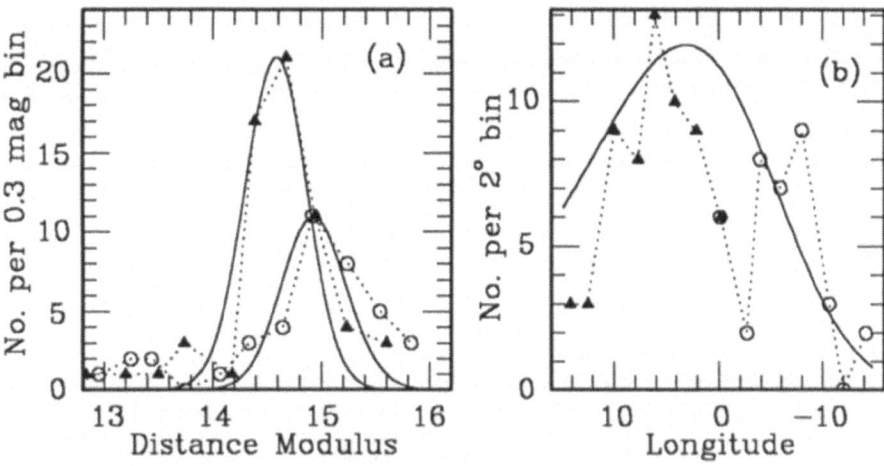

Fig. 3. The same as Fig. 2, together with one of the best fitting models (prolate ellipsoid).

$b = 0$, such that $0< \theta <90$ implies that the major axis lies in the first quadrant.

Figure 3 shows an example of one of the best fitting model Bulges. In this and all other fits the model has been convolved with a $\sigma = 0.2$ mag gaussian distribution to allow for the observational error on the distance moduli. The illustrated model, which provides a reasonable fit to the data, is a prolate spheroid with $x_o = 760$ pc, $y_o = z_o = 190$ pc, and $\theta = 45°$. The distance to the galactic centre ($R_0$) is $R_0 = 9.1$ Kpc. Given the large number of parameters we are fitting to a statistically small number of stars it is not possible to establish a unique solution. It is however possible to derive some insight into the range of plausible parameters.

Note that our equation (1) models an ellipsoidal stellar distribution which is distinctly different from the rhomboidal distribution used by Blitz & Spergel (1991 their equation (1)) to model the 2 $\mu$m emission from the Bulge.

Harmon & Gilmore (1988) derive a scale height of 375 pc, if $R_0 = 8$ Kpc, for colour selected IRAS sources in the Bulge. The sources they discuss are very similar to those we are considering here (although they are drawn from a much larger volume of the Bulge) and we would therefore expect to derive a comparable scale height. The progenitors of the Miras in the Bulge are the late-M stars and they have scale heights of between 260 and 300 pc (Blanco & Terndrup 1989). These values cannot be compared directly with the scale lengths used in equation (1), because of the tilt of the Bulge to the line of sight. They must be compared with a surface density scale height ($h_o$) which can be calculated for each model. The model shown in Fig. 3 has $h_o = 212$

pc. It is not possible to get a good fit to Fig. 2 with $h_o > 290$ pc.

Given the slightly different nature of the various samples the agreement is satisfactory. It is important to recall that we are modelling two narrow strips through the Bulge (see Fig. 1) which are only about one scale height thick and are of the order of five to six scale heights away from the centre. Bearing this caveat in mind the great power of this approach is that we can deal with individual distances rather than surface intensities or numbers per unit area.

Binney *et al.* (1991) argue that the inclination of the bar-shaped Bulge must be $\theta = 16\pm2°$ in order to explain the dynamics of the galactic centre gas. It is possible to get a good fit to Fig. 2a with this low value of $\theta$ and a reasonable scale height of $z_0 = 150$ pc. However this model results in a very narrow distribution in longitude which is inconsistent with Fig 2b. Of course it is not possible to reproduce the dip in the middle of the longitude distribution (Fig. 2b) at all with a triaxial ellipsoidal Bulge. This dip is what one might expect from an X- or peanut-shaped Bulge.

Figure 4 shows the result of fitting an X-shaped Bulge to the data. This was modelled using two identical triaxial ellipsoids inclined at an angle $\alpha = 60°$, to each other and at an angle $\theta = 45°$ to the line of sight. This does fit the data rather better than the simple ellipsoid. However an improvement of the fit is to be expected purely because we have increased the number of parameters. Note that similar structure is *not* seen in the longitude distribution of late M-stars at $b = -6$ (Blanco & Terndrup 1989). Nevertheless the suggestion of an X-like structure is interesting and worth further investigation, but in view of the limited data it is premature to draw any further conclusions here. Other evidence for the X-shaped nature of the Galactic Bulge has been summarized by Freeman (1990).

It is also possible to fit Fig. 2 with a power law number density distribution, e.g. of the kind used by Blanco & Terndrup (1989) to fit the M-star data. However a rather high power, about 6, is necessary to do this. Such a high value would imply an unrealistically large number of stars at smaller galactic latitudes.

## 4. Rotation of the Bulge

Menzies (1990) discussed kinematic data for 26 IRAS Miras from the Whitelock *et al.* sample. These stars have a typical Bulge velocity dispersion ($76\pm11$ km s$^{-1}$) and show clear evidence for rotation, with a velocity gradient of $9.8\pm1.9$ km s$^{-1}$ deg$^{-1}$. This corresponds to a rotation frequency of $\omega \sim 70$ km s$^{-1}$ kpc$^{-1}$, in good agreement with that predicted for the bar by Binney *et al.* (1991) in the absence of significant streaming in the bar's frame of reference.

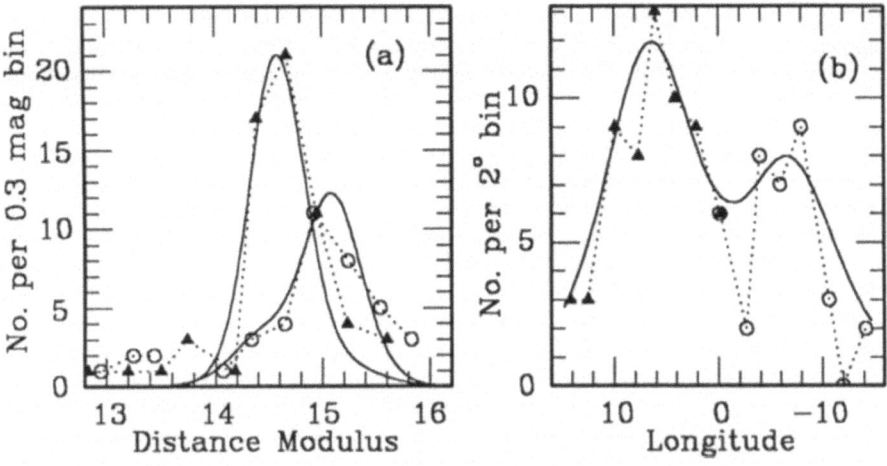

Fig. 4. The same as Fig. 2, with an 'X-shaped' Bulge model

## 5. Conclusion

The number distribution of Miras in the Bulge as a function of distance modulus indicates that the half of the Bulge which is at positive galactic longitudes is closer to us than the half at negative longitudes. The observed stellar distribution can be modelled with a prolate ellipsoidal distribution of stars which we refer to as a 'bar'. This bar is tilted at roughly 45° to the line of sight. This conclusion is consistent with that of Blitz & Spergel (1991).

## Acknowledgements

We are grateful to Michael Feast, John Menzies and Kaz Sekiguchi for helpful discussion.

## References

Binney, J., Gerhard, O. E., Stark, A. A., Bally, J. & Uchida, K. I.: 1991. *Mon. Not. R. astr. Soc.*, **252**, 210.

Blanco, V. M.: 1988. *Astronom. J.*, **95**, 1400.

Blanco, V. M. & Terndrup, D. M.: 1989. *Astronom. J.*, **98**, 843.

Blitz, L. & Spergel, D. N.: 1991. *Astrophys. J.*, in press.

de Vaucouleurs, G.: 1964. In *The Galaxy and the Magellanic Clouds*, IAU Sym. 20, Eds. F. J. Kerr & A. W. Rodgers, p. 195.

Feast, M. W.: 1963. *Mon. Not. R. astr. Soc.*, **125**, 367.

Feast, M. W. & Whitelock, P. A.: 1987. In *Late Stages of Stellar Evolution*, Eds. S. Kwok & S. R. Pottasch, p 33. Dordrecht, Reidel.

Feast, M. W. & Whitelock, P. A.: 1990. In *Bulges of Galaxies*, ESO-CTIO Workshop, p. 3.

Freeman, K. C.: 1990. In *Dynamics and Interactions of Galaxies*, Ed. R. Wielen, p.36.

Gerhard, O. E. & Vietri, M.: 1986. *Mon. Not. R. astr. Soc.*, **223**, 377.

Harmon, H. & Gilmore, G.: 1988. *Mon. Not. R. astr. Soc.*, **235**, 1025.

IRAS Point Source Catalogue: 1985. US Government Publication Office.

Matsumoto, T., Hayakawa, S., Koizumi, H., Murakami, H., Uyama, K., Yamagami, T. & Thomas, J. A.: 1982. In *The Galactic Centre*, Eds. G. R. Riegler & R. D. Blandford, p. 48.

Menzies, J. W.: 1990. In *Bulges of Galaxies*, ESO-CTIO Workshop, p. 115.

Nakada, Y., Deguchi, S., Hashimoto, O., Izumiura, H., Onaka, T., Sekiguchi, K. & Yamamura, I.: 1991. *Nature*, **353**,140.

Terndrup, D. M.: 1988. *Astronom. J.*, **96**, 884.

Whitelock, P. A.: 1990. In: *Confrontation of Stellar Evolution and Pulsation Theory*, Eds. C. Cacciari & G. Clementini, PASP Conf. Ser. p. 365.

Whitelock, P. A., Feast, M. W. & Catchpole, R. M.: 1991, *Mon. Not. R. astr. Soc.*, **248**, 276.

# HI IN THE INNER GALAXY

## HARVEY S. LISZT

*National Radio Astronomy Observatory,*
*520 Edgemont Road, Charlottesville VA USA 22903-2475*

**Abstract.** I consider several aspects of the Galactic HI distribution, kinematics, and morphology at $R < R_0$, in particular; the rotation curve of the inner few kpc; the variation of HI across the inner disks of the Milky Way and other galaxies; the amount of HI inferred in the disk (with implications for the $CO-H_2$ conversion factor); and lastly, the physical disposition of neutral atomic gas and some consequences of the required structure of HI 'clouds' for optical and *uv* absorption-line profile analysis.

## 1. Introduction

This paper is the result of a talk entitled 'HI In The Galaxy'. In fact, such an all-inclusive title applies to several insightful and thought-provoking recent reviews by Kulkarni and Heiles (1987), Burton (1988), Heiles and Kulkarni (1988) and Dickey and Lockman (1990), but not so generally to this text (or to the talk on which it is loosely based). Instead I intend to provide here something of a limited guide to several specific aspects of HI studies and related issues of Galactic structure, with two intended purposes.

My first intention is to acquaint the audience of consumers of HI-related results–those who merely need the answers–with some of the difficulties and inconsistencies which arise in determining them. Beyond that, I wish to clean up some loose odds and ends, making some new associations and drawing some new conclusions which are grounded in the oldest and very most fundamental properties of HI observations and observational programs. As subject matter, I will treat areas which are the bread and butter of HI studies, namely, the rotation curve of the inner Galaxy (Section 2); the distribution and quantity of atomic gas across the disk of the inner Milky Way (Section 3), and the volume and surface density of HI near the Sun and at $R = R_0$ (Section 3.4). In Section 4, I discuss the filling factor and physical disposition of the HI, showing that the observed HI line profiles have direct consequences for interpretation of the same gas observed optically, *i.e.* for optical/uv absorption line analyses of diffuse clouds.

Throughout this paper the values $R_0 = 8.5$kpc, $\Theta_0 = 220$ km s$^{-1}$ are employed.

## 2. The Rotation Curve of the Inner Milky Way

For about a decade it has been increasingly apparent that the 'rotation curve' of the Milky Way as usually presented is wrong or possibly even irrelevant at Galactic radii of a few kpc. Yet this idea has been slow to permeate the

111

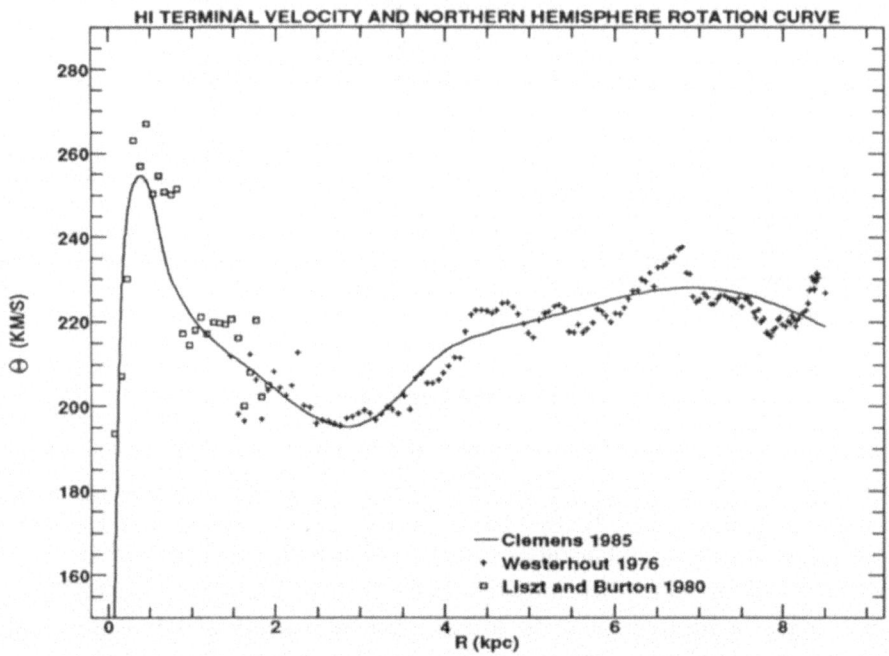

Fig. 1. HI terminal velocity observed at $l > 0$, from data of Westerhout (private communication) and Liszt and Burton (1980). The polynomial rotation curve of Clemens (1985), derived chiefly from CO observations, is shown superposed.

general community, with the result that detailed mass models of the Galaxy have been explicitly constructed to suit inappropriate constraints.

In the disk of the Milky Way, which in HI really means at about $R \geq 0.45R_0 = 3.8$kpc, there is little doubt that the motions of interstellar HI gas are, in declining order of speed, 1) circulation at about 220 km s$^{-1}$; 2) occasional spiral streaming motions up to 15-20 km s$^{-1}$; and 3) ubiquitous random motions of about 6 km s$^{-1}$. The circulation may not be purely circular, with $\approx 15$ km s$^{-1}$ radial motion due to the inherent ellipticity of gas orbits (Blitz and Spergel 1991), but pretty generally, and probably safely, we interpret the run of HI terminal velocity (unbiased maximum velocity) along the line of sight as reflecting the equilibrium circulation speed and interior gravitational mass as a function of radius.

The rotation curve derived from observations of CO in the usual way by Clemens (1985) (see also Burton and Gordon 1978) is shown in Figure 1. Inside the disk, the shape of our supposed rotation curve is quite strange (although there is a nearby galactic neighbor which, rather uncomfortably, exhibits a similar phenomenon), something the reader is invited to confirm by looking either at the many optical rotation curves compiled by Rubin

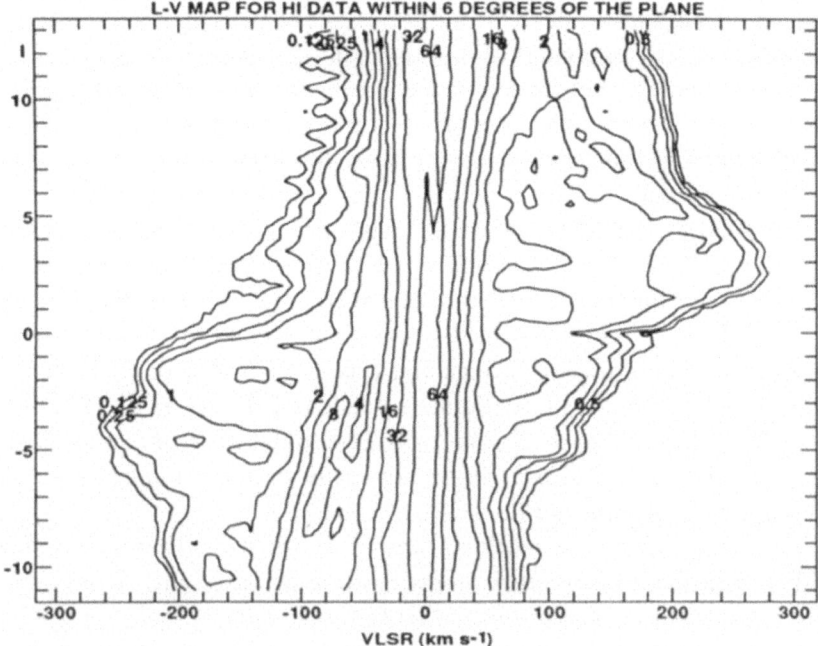

Fig. 2. *l*-v diagram of HI emission summed over $|b| \le 6°$ using data of Liszt and Burton (1980). Contours of antenna temperature move up in steps of 2 from 0.125 K. Note the continuity of the extrema of the envelope of emission on either side of $l = 0$.

(1983) or the recent compendium of extended HI rotation curves by Cassertano and van Gorkom (1991). The rapid rise, subsequent dip, and resumption are unusual, to say the least, in the galaxies represented in these references. A somewhat similar shape in NGC3200 was pointed out by Rubin (1983), but the structure in that object is not nearly as compact.

Figure 2 shows some of the data on which the inner-Galaxy HI rotation curve might be based, but now presented in such a way as to highlight the impermissibility of the usual practices; this point was made explicitly in Figures 5 of Burton and Liszt (1978) and Liszt and Burton (1980), who demonstrated the tilt of the inner-Galaxy gas layer, but that aspect has been suppressed here by the simple tactic of integrating across the Galactic plane at each longitude. The point to be made is that although there is nothing to stop us from deriving and plotting the positive-velocity terminal velocity at $l > 0$ or the negative-velocity terminal velocity at $l < 0$ and labelling it as a maximum rotation velocity, it would be entirely nugatory to do so.

Consider the extreme positive-velocity envelope in Figure 2, which seems not in the least to 'know' or 'care' where it crosses the artifactual but highly significant half-plane at $l = 0$ (artifactual because its precise location is an accident of the location the Sun). At $l < 0$, a gas parcel acquires its perceived speed (algebraically) in the sense (non-circular - |circular|); at $l > 0$, the circular component of motion changes sign and the speed is (non-circular + |circular|). Yet the velocity envelope is continuous, and clearly the contribution of the circular component of motion is not so large as to cause a discontinuity in that envelope, only a gradient, even though the velocity is high at $l > 0$ and $v > 0$.

The HI data must be therefore be construed as reflecting a combination of circular and non-circular motions which is so well-balanced that even material at the highest rotationally-permitted velocities is made up of a mixture of both in which neither is negligible; this is the point of a small but steady stream of papers (Cohen and Davies 1976, Burton and Liszt 1978 and Liszt and Burton 1980, Heiligmann 1983, and Binney *et al.* 1991) which have deduced the existence of a bar or triaxial potential from analysis of gas motions in the inner Galaxy. There are many important inferences which can be drawn from this work.

• The inner-Galaxy gas motion always deviates strongly from circularity.

• The rotational component of gas motion increases with R much more slowly than does the usually-plotted HI terminal velocity, because the latter includes contributions from non-circular motions which are large at small R and persist for several kpc.

• Even when the rotational component of motion can be isolated, and its variation with R traced, this is of little use if the gas is not in pure rotation and the circular component of motion is neither dominant nor easily related to the total gravitational potential.

As a last word in this matter, I note that there is yet another sense in which difficulty will arise in deriving *the* rotation curve in the inner regions of the Galaxy from observations of CO or HI, because there can be more than one of them even for the same species of gas. Consider the 2pc ring at the center of the Galaxy, for which the motion is shown in HI and CO by Liszt *et al.* (1985) and again very nicely in CO by Fukui and Churchwell(1987); with the possibility of a 50 km s$^{-1}$ non-circular motion (also seen in recombination lines by Goss *et al.* 1990), the dominant motion is rotation at 80–110 km s$^{-1}$. However, there is also a more widely distributed component of CO seen in the Galactic plane with a strong rotation signature (the molecular analog of the classical HI rotating nuclear disk; see Liszt and Burton 1978) within which the velocity reaches 100 km s$^{-1}$ only at $|l| \geq 0.5°$ ($R_0 \geq 75$pc projected distance).

The moral of this story is that the neutral gas reveals a wide variety of interesting and poorly-understood phenomena, but does not at present do a creditable job of reflecting the distribution of stellar mass and gravitational potential. The nearby galaxy alluded to above, which has a similarly unusual shape to its inner-galaxy rotation curve, at least in ionized gas, is M31.

Ciardullo *et al.* (1988) attribute the behaviour in M31 to outflow induced by stellar winds. In the Milky Way pure outflow would involve a much greater expenditure of momentum, simply because there is so much neutral gas with non-circular motions, and this explanation probably does not apply.

## 3. The Atomic Gas Distribution At $R < R_0$

The amount of gas in the Milky Way is clearly of great interest, as is the manner in which that gas is aportioned among atomic, molecular, and ionized components, between cloud and intercloud components, and between constituents of different scale heights. For instance, it is widely believed that the total amount of neutral gas increases moving inward of the Solar Circle, as there are many reasons to expect from observations of other galaxies (even disregarding those of our own!), but the amount of HI seen in the Galactic equator is apparently constant (Burton 1988, and below). Thus the ratio of molecular to atomic gas should increase moving inward, perhaps by a substantial amount. If the total amount of neutral gas near the Solar Circle is roughly equally divided between molecular and atomic gases–the prevailing view–the ratio of molecular to atomic gases in the inner regions of the Galaxy ($R \geq 0.5R_0$) must be heavily molecular. However, the latter fact is not universally accepted, and claims of a small atomic/molecular gas ratio in the inner Galaxy have generally been denigrated in the last half decade.

Our ability to trace molecular gas across the disk of the Milky Way has been the subject of no small amount of controversy, because so much faith must be placed in observations of carbon monoxide, whose integrated intensity is taken to be proportional to the amount of molecular gas under a wide range of (often poorly-understood) conditions. To be sure, the amount of molecular gas interior to the Solar Circle, inferred by various observers, has varied much more substantially than that of HI (with a clear downward trend). Although there is little controversy surrounding quoted values of the HI mass of the Milky Way, it is considerably less certain than might otherwise be believed. The density of HI gas in the Galactic plane, derived from HI observations by real aficionados of 21cm studies, varies typically by 50-100%.

### 3.1. HI as a uni-component gas in the plane of the Milky Way

It has been recognized at least since the work of Heeschen (1954) that the HI observed at low Galactic latitudes is inhomogeneous, an admixture of colder and warmer material. Of course this is a gross understatment of the complexity of the atomic gas, but the fact remains that, having failed to formulate a clear and distinct idea of the distribution and topology of these

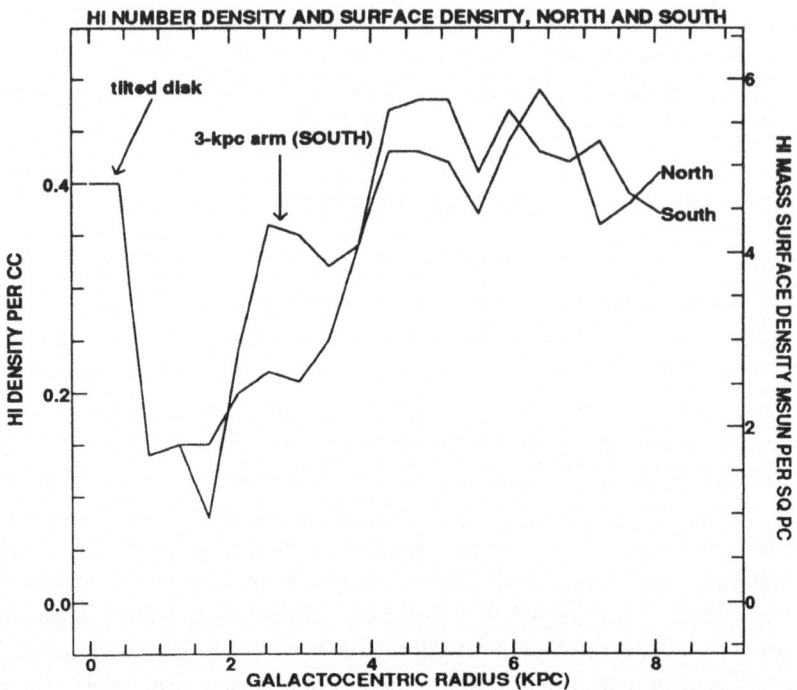

Fig. 3. HI density derived from observations at $b = 0°$ as in Burton and Gordon (1978), but using modern Galactic constants. Data are as in Figure 1, supplemented by observations of Kerr, Kerr, and Bowers (private communication) in the South. The right-hand scale of mass surface density is the standard fudge based on normalizing results to the column density observed locally at high latitudes

sub-constituents of the (atomic) ISM, we cannot entirely model their emission or derive directly the quantity of gas. The simplest approach to derivation of the HI density, whereby the HI is viewed as a fully volume-filling gas of uniform temperature (which is quite clearly 'wrong'), is still the only method which has been formulated and employed self-consistently. In spite of its simplicity it is surprisingly robust in the sense of reproducing the two most general aspects of the actual HI distribution, namely the integrated emission and absorption strengths.

¿From observations of HI absorption toward pulsars (Frail *et al.* 1991) it is found that the total integrated optical depth per kpc of path (referred to the Galactic plane) is about 5 K km s$^{-1}$ kpc$^{-1}$. It follows then, that on average an optically thin gas column 1 kpc long at temperature $T_{\rm sp}$ should produce $5\,T_{\rm sp}$ K km s$^{-1}$ integrated intensity, which translates directly into a column density, and hence (over a kpc) a density. With a value $T_{\rm sp} = 135$ K, it happens that the required value (density) is 0.41 HI cm$^{-3}$. Perhaps surprisingly, this is the value which is found in the plane of the Galaxy, interior to the Solar Circle, from modelling of the *emission* using a one-component gas. The spin temperature of 135 K is the highest brightness temperature observed at low latitudes.

This simple coincidence of reproducing the gross opacity *and* brightness at a single temperature strongly suggests that the emission measurements cannot be misleading us too seriously when the gas quantity is derived from them. It would be hard to imagine a hidden component of HI lurking in the inner Galaxy which contains a substantial amount of material without affecting one or the other (emission or absorption) very substantially. Nonetheless, we will see that the amount of HI gas directly derived across the Galaxy is somewhat too small to account for the amount of gas seen overhead near the Sun.

Figure 3 shows the HI density derived in the Galactic equator, very much as in Burton (1988), but now using more recent values of $R_0$ and $\Theta_0$. The emission measurements in principle determine the HI surface density at any radius (although they do not apparently get it right just off the bat); with smaller $R_0$ the inferred scale height is smaller but $n_{HI}$ is bigger. To produce this diagram, I used the rotation curves of Clemens (1985) and Alvarez *et al.* (1990) in the Northern and Southern Hemispheres, respectively. The correct location of the so-called 3-kpc arm gas is uncertain for reasons described in the previous section.

## 3.2. SURFACE DENSITY, GALACTIC HI MASS, AND UNCERTAINTY IN MID-PLANE DENSITY

The right-hand scale of HI mass surface density in Figure 3 is pretty much of a fudge based on normalizing all results to the value of the HI mass surface density observed at high latitudes in the solar vicinity, $6 \times 10^{20} \text{cm}^{-2}$ (Dickey and Lockman 1990); it cannot be derived directly by measuring the vertical (z-) distribution from the same data used to derive the density. The z-scale height derived from Figure 3 by taking the ratio $N_{HI}/n_{HI}$ is unphysical or wrong (or both).

To substantiate this claim, the reader can compare our plot of $n_{HI}(R)$ with two recent derivations of the vertical distribution of the HI number density in Lockman (1984; also quoted in Burton 1988) and Dickey and Lockman (1990). In the earlier of these, the mid-plane density is given as 0.3 HI cm$^{-3}$, somewhat lower than our value of 0.41 cm$^{-3}$ because no correction was made for saturation, because a larger telescope beam was employed and because the older value $R_0 = 10$kpc was used. In the later work, the midplane density is cited as 0.6 cm$^{-3}$ (for $R_0 = 8.5$kpc). This value was achieved by *adjusting* the mid-plane density of the thinnest, physically coldest HI component (corresponding to 'clouds') by that amount necessary to achieve the correct area under the curve–the locally determined HI surface density. However, the total midplane density in Dickey and Lockman will

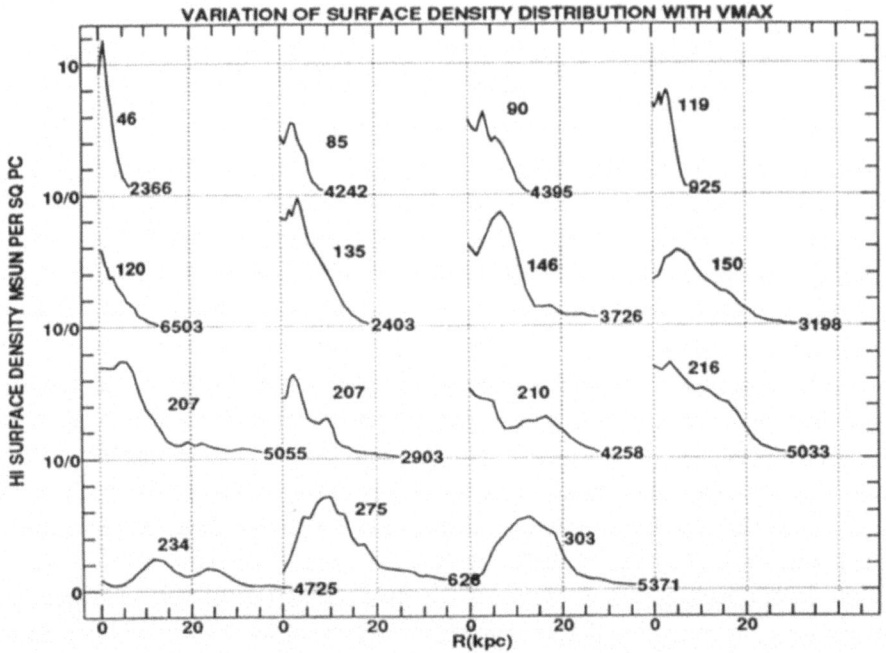

Fig. 4. Variation of HI radial surface density distribution with maximum rotation velocity for 15 systems studies by Wevers (1984). The NGC number for each system is shown at the tail of its curve; $V_{max}$ is shown inset in each panel.

not reproduce the observations because either the integrated emission or absorption will be too large. Their upward adjustment is essentially just the same fudge used here to produce the right-hand scale in Figure 3.

### 3.3. VARIATION ACROSS THE DISK; EVOLUTION WITH HUBBLE TYPE

It is by now well known that there is a type and luminosity sequence of shape and height of the rotation curves of galaxies (see Rubin 1983 and Cassertano and van Gorkom 1991); earlier and brighter spirals have rotation curves which rise more sharply near the nucleus, and to higher peaks. There is also a correlation between type or maximum rotation velocity and shape of the HI distribution. Later-type systems with lower maximum rotation speeds have their HI more centrally concentrated while a distinct HI disk occurs only midway through the Hubble sequence of spirals. In general, systems in which the density of HI is high in the nucleus do not have a distinct HI annulus in the disk (*i.e.* they do not have a gap between the disk and nucleus as the Milky Way does) but show only a gradual decline with radius. Like M31, sufficiently early spirals having a clear HI disk or large ring about the nucleus lack neutral atomic gas in their inner regions. This situation is illustrated in Figure 4 for 15 galaxies mapped in HI by Wevers (1984).

The HI distribution of the Milky Way is mildly unusual in this context for it neither falls smoothly and continuously with increasing R nor forms a disk with a complete central hole. Instead, in our Galaxy there is a fairly high density of HI both at R < 1 kpc and R > 4 kpc. If an explanation is sought in terms of errors in interpreting the Milky Way, I reiterate the difficulties that await anyone using the gas rotation curve inside the proper gas disk, and the inconsistencies that arise in associating the measured number and surface densities. For the external systems, there are often problems of spatial resolution, although an annular hole 2 kpc across would be hard to miss in many galaxies.

The subject of variation across the disk has had some currency in CO studies, where we have subjected ourselves (perhaps rightly) to a degree of agony in debating whether the integrated CO emissivity reflects truly the variation of the molecular gas quantity in the presence of such galactic effects as changes in metallicity. However, there is very little direct evidence for variation in the internal properties of the gas (or gas clouds) across the disk, even in CO (however see Liszt *et al.* 1984). In HI, even some expected variations across the disk are not clearly present. For instance, the residual amounts of HI present in molecular clouds–which may achieve high HI opacity due to low temperature–are expected to yield an increased integrated optical depth per kpc of path in the inner Galaxy where the quantity of molecular clouds is very much enhanced (*ibid*).

## 3.4. HI AND TOTAL GAS DENSITIES NEAR THE SUN

One of the most commonly cited results concerning the local interstellar medium is that of Munch (1952), as quoted by Spitzer (1978), whereby the statistics of extinction toward a very large number of A-stars yield the local mean extinction per kpc of path in the Galactic plane and the spatial frequency of absorbing gas parcels (6 $kpc^{-1}$). The extinction, converted in modern fashion to a gas column density yields a mean density of 1.2 H-nuclei $cm^{-3}$, much larger than the values 0.3, 0.4, and 0.6 HI $cm^{-3}$ which have been 'derived' from HI as discussed above. Munch's results must have been at least slightly biased away from the darkest and densest clouds. Alternatively, some small part of the extinction may perhaps be associated with hotter, ionized material.

The difference between the total and HI local gas densities at z = 0, which must be *at least* 0.6 H-nuclei $cm^{-3}$, is in the form of molecular material. The existence of an independent determination of the local density of such matter implies an estimate of the conversion factor between CO intensity and hydrogen column density. CO surveys report an emissivity of 4 K km $s^{-1}$ $kpc^{-1}$ at the Solar Circle, implying a value of $2.3 \times 10^{20}$ $H_2$ (K km $s^{-1}$)$^{-1}$.

This is, in fact, about the value which is used to derive the molecular gas density generally.

One ramification of this discussion is that the CO-$H_2$ conversion factor cannot be substantially lower than the currently used value without placing unreasonable demands on the atomic component of the local ISM. The HI data simply do not support the very high local atomic gas densities implied by such too-small conversions.

## 4. The Physical State of HI in the disk of the Milky Way

I showed above that a simple *ansatz* reproduces the gross character of HI absorption and emission in the Galactic equator, albeit with the unhappy consequence that a fudge is necessary to reconcile the quantities of HI observed in the plane ($n_{HI}$) and across it ($\sigma_{HI}$ or $N_{HI}$). The discrepancy points out the fact that the disposition of Galactic HI is only marginally understood.

The simple conception of Galactic HI as a uniform, 135K, volume-filling gas has undergone several important stages of modification as observational aspects became newly available and were subsequently assimilated and explained theoretically. The first of these was the raisin-pudding model and two-phase interstellar medium which in turn was employed successfully in a simulation by Baker and Burton (1975) to explain the spatial fluctuations of Galactic HI emission. In this picture, 80 K clouds of size a few pc, having a mean free path of 1/3 kpc and volume filling factor $f3$ of order a few percent, are immersed in a pervasive, warm (8000K) neutral intercloud medium with a somewhat larger scale height (100 vs 70 pc for our value of $R_0$) and intrinsic velocity dispersion.

The two-phase model was challenged theoretically in the 1970's as it was realized that a substantial fraction of the volume of the local Galactic disk would be occupied by hot ($10^6$K), low-density gas arising from the expansion and interaction of supernova remnants. In turn, a series of HI experiments begun by E. Salpeter, Y. Terzian, and their students at Arecibo pointed out several failings of the strictest interpretation of the two-phase model and found that some fraction of the hot intercloud gas must be physically associated with the colder absorbing gas parcels ('clouds'), not simply surrounding them (for the latest installment in that series, see Colgan *et al.* 1986).

Most recently, we have come to realize that the absorbing gas parcels known frequently as clouds are not little bullets meandering aimlessly in space, but have been coerced, under external influences, into larger scale structures like shells and loops, which fill only a portion of the disk.

## 4.1. VOLUME AND AREA FILLING FACTORS FOR HI

The observable HI gas fills a fraction of space which is, at present, very uncertain. As I summarized elsewhere at this meeting (Liszt 1992), observations of other components of the ISM–the warm $(10^4 K)$ ionized gas providing the pulsar dispersion measures and diffuse $H\alpha$ emission, and the hot $(10^6 K)$ OVI-bearing gas–imply that the volume filling factor of the HI intercloud gas is no more than 0.7. For a long time, it was inferred from the ubiquity of HI emission that the HI filled essentially all of the Galactic disk. Although there is no known line of sight with a column density of HI below $4.4 \pm 0.5 \times 10^{19}$ cm$^{-2}$, even after the most exacting correction for confusion (Lockman et al. 1986), it still does not follow that the HI necessarily fills more than about 30% of the disk (at z = 0).

To see this, consider for some class i of interstellar object the spatial mean free path $\lambda_i$ whose inverse $\nu_i$ is the frequency with which class members are encountered in a random line of sight at z = 0. For interstellar clouds studied optically and at $\lambda 21$cm, $\nu_i = 1/\lambda_i = 6 - 10$kpc$^{-1}$. This is the usual definition in which $\lambda_i = (n_i\sigma_i)^{-1}$, where $n_i$ is the spatial number density and $\sigma_i$ is the cross-section. The fraction of the volume occupied by the class i is $f3_i = d_i/\lambda_i$ within a factor of order unity (if i represents spheres, $d_i$ is the diameter and the factor is 2/3). Identical spheres are fully packed into a volume when $f3 = \pi/6$.

This stands in contrast to the covering or area filling factor $f2$, which is related to the probability that a line of sight will cross the Galaxy perpendicular to the Galactic plane and not encounter an instance of class i; this probability is 1-$\exp(-H/\lambda_i)$ where H is the effective thickness of the Galactic layer. I define the covering factor, which can exceed unity, as $f2_i = H/\lambda_i = (H/d_i)f3_i$. Typically $H/d_i \gg 1$ because we consider objects many of which fit comfortably within the Milky Way, and so it is that even a moderate volume filling factor can be consistent with a very high probability of occurence along every line of sight. For the absorbing material of Galactic HI, with $\lambda = 100$ pc and $f3 \approx 5\%$, a seemingly ubiquitous intercloud medium could be constructed by putting a warm envelope around each absorbing gas packet. With a ratio of characteristic sizes of (say) 2.5:1, the warm gas would have a mean free path of 15 pc and a volume filling factor of 30%.

## 4.2. RAMIFICATIONS OF A SMALL VOLUME FILLING FACTOR FOR HI

The uncertain but small (non-unity) volume filling factor of interstellar HI and the observational inference that cold and warm HI are physically associated on a 'cloud-by-cloud' basis combine to have important consequences for studies of HI. Just as importantly, however, they dictate certain very significant aspects of diffuse cloud structure whose effects should be appreciated by a broader audience.

The essential remark is the cliché that HI emission is broader than HI absorption. This is true in two senses. The first, that HI emission fills the allowable velocity space within a line profile so much more fully than HI absorption, is the raisin-pudding model. The second sense is that HI emission and absorption, when they *can* be associated within a single kinematic feature, unerringly exhibit the property that the emission line is broader than its underlying optical depth profile. Any such feature must be heavily structured internally; the only way to produce dissimilar emission and absorption profiles is to blend hot and cold or weakly and strongly absorbing gas. This is the three-phase model of McKee and Ostriker (1977), albeit with heavy modifications to accomodate the real world of HI profiles. As was pointed out earlier, it is not sufficient simply to put an entirely distinct hot envelope around a cold core (Liszt 1983). There must be a continuous change of temperature in the gas or the HI line profiles do not resemble the observations in the least.

This structuring was remarked by Radakrishnan *et al.* (1972) in their classic interferometer study of HI absorption, and reiterated in consideration of the single-dish analog of that work as performed at Arecibo (Liszt 1983). But in neither case were the wider implications of that remark made clear. Diffuse clouds have a characteristic structure whose influence will be felt just as strongly in optical curve-of-growth or profile analysis.

## 4.3. A HEURISTIC MODEL OF λ21CM HI LINES

A packet of gas with column density $N_{HI}$, temperature $T_{sp}$, and gaussian velocity dispersion $\sigma_v$ has an opacity to λ21cm radiation

$$\tau(v) \propto \frac{N_{HI}(v)}{\sigma_v T_{sp}} \tag{1a}$$

and, in the presence of a background radiation field characterized by brightness temperature $T_{bg}$, produces an outgoing emission contribution

$$T_b(v) = T_{sp}(1 - e^{-\tau(v)}) + T_{bg}e^{-\tau(v)}. \tag{1b}$$

However, when $T_b(v)$ and $(1 - e^{-\tau(v)})$ are measured across the profile of real (that is to say, observed) clouds, it is found that the ratio $T'_{sp}(v) = T_b(v)/(1 - e^{-\tau(v)})$ increases away from the line center. This is not representative of experimental errors but of blending of hotter and colder material along the line of sight. Some of the hotter material arises external to the 'cloud' in question, and even a uniform cloud seen along with the 'inter-cloud' medium will exhibit such behaviour to some small degree. However, the observed effect is much more substantial than expected from a two-phase medium because the typical FWHM measured in emission exceeds that in absorption by 25-50% or more.

To devise a scheme in which a cloud can produce a profile $T_b(v)$ wider than $(1 - e^{-\tau(v)})$ one might attempt to lower the opacity at the HWHM point in the profile while maintaining the brightness temperature there. For HI this is done straightforwardly by increasing the contribution of hotter gas at the HWHM relative to that at the line center, which in turn occurs if the cloud temperature increases outward and the line width increases with temperature. In this way we are led to consider a velocity dispersion $\sigma_v$ expressed as

$$\sigma_v^2 = \sigma_t^2 + \frac{T_k}{121\mathrm{K}} \quad (\mathrm{km} \quad \mathrm{s}^{-1})^2 \tag{2a}$$

where 121 K is the kinetic temperature at which the velocity dispersion is 1 km s$^{-1}$ for purely thermal motion. The quantity $\sigma_t$ is adjustable but independent on $T_k$ and is employed to reproduce the fact that linewidths in HI are always somewhat supersonic in cold material; typically, for $T_{sp}' = 80\mathrm{K}$ one sees $\sigma_v = 1.3$ km s$^{-1}$, implying $\sigma_t = 1$ km s$^{-1}$.

Hotter gas should also be more tenuous. To relate $n_{HI}$ and $T_k$ we will assume turbulent pressure equilibrium, i.e.,

$$n_{HI}\sigma_v^2 = const. \tag{2b}$$

which is equivalent to $n_{HI}T_k = const.$ for small $\sigma_t$ or large $T_k$, but the core value of $T_k$ is fairly low here, 40 K, and the added contribution from $\sigma_t$ ensures the presence of a small core of cold gas (supported by turbulent rather than thermal motion).

At this point, all quantities necessary to calculate $T_b(v)$ are parametrized in terms of $T_k = T_{sp}$. If the transition between cold and hot gas is too abrupt the broad-lined hot gas and the narrow-lined cold material produce largely disjoint, easily separable HI line profiles Liszt (1983). What is needed heuristically is a gentler and more continuous structure allowing substantial gas quantities at intermediate linewidths and temperatures. Transitions between temperatures $T_0 < T_1$ were parametrized as

$$T(r) = T_1 + (T_0 - T_1)\exp(-qr^2) \tag{3}$$

where r is fractional cloud radius and $q > 0$ provides a means to partition the cloud material between a cool core and hot outer envelope. The assumed geometry is spherical, but this is of no consequence here. The lowest $T_k$ in a model must be taken somewhat below the typical values measured for $T_{sp}'$ at the centers of HI components and the highest must be at or above the lower limits derived for $T_{sp}'$ in the 'intercloud' gas seen in the line wings.

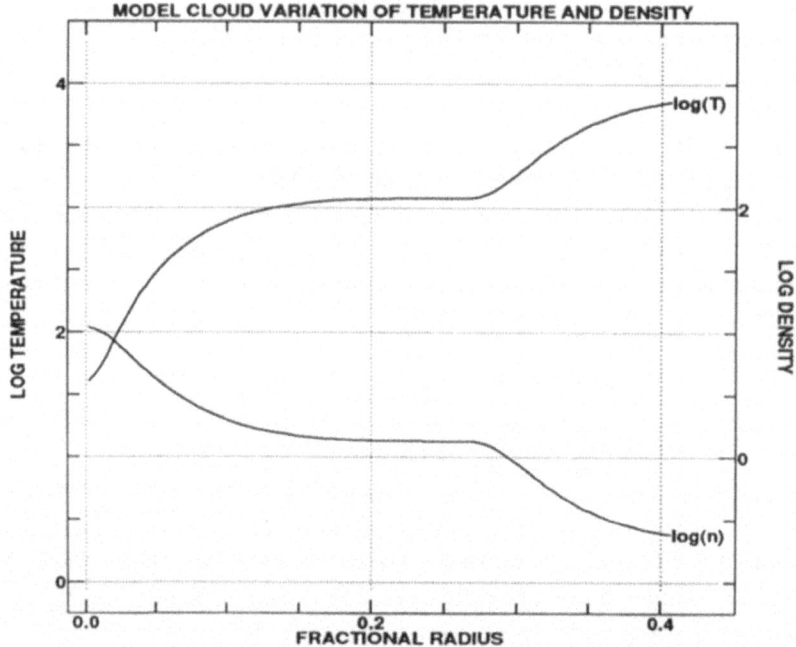

Fig. 5. Variation of density and temperature over the inner regions of a model having temperature zones at 40, 1200, and 8200 K and total column density $N_{HI} = 3 \times 10^{20}$ HI cm$^{-2}$.

To complete the specification of a cloud model it is necessary to choose the total column density and physical size. These are available from observation and typical values are $N_{HI} = 3 \times 10^{20}$ cm s$^{-1}$ and radius 5 pc (see Baker and Burton 1975). Shown in Figure 5 are the run of density and temperature over the inner regions of such a model having $\sigma_t = 1$ km s$^{-1}$ and plateaus with temperature 40, 1200, and 8200 K.

Figure 6 shows the profiles of brightness temperature and optical depth from the model. The former may be decomposed into three gaussian components having widths (FWHM) of 3.4, 7.0, and 17.2 km s$^{-1}$, with areas in the proportion 9:8:4. The opacity is fit by two gaussians having widths of 3.1 and 7.2 km s$^{-1}$ with areas in the proportion 9:1.

## 4.4. Optical/UV Absorption Line Profiles From HI Clouds

The previous subsection mostly reiterates the final discussion in Liszt (1983), wherein it was argued that clouds so structured could reproduce the detailed appearance of observed $\lambda$21cm line profiles, and certain statistical inferences drawn therefrom, while maintaining the semblance of an intercloud medium with a volume filling factor substantially below unity. Here I wish to go

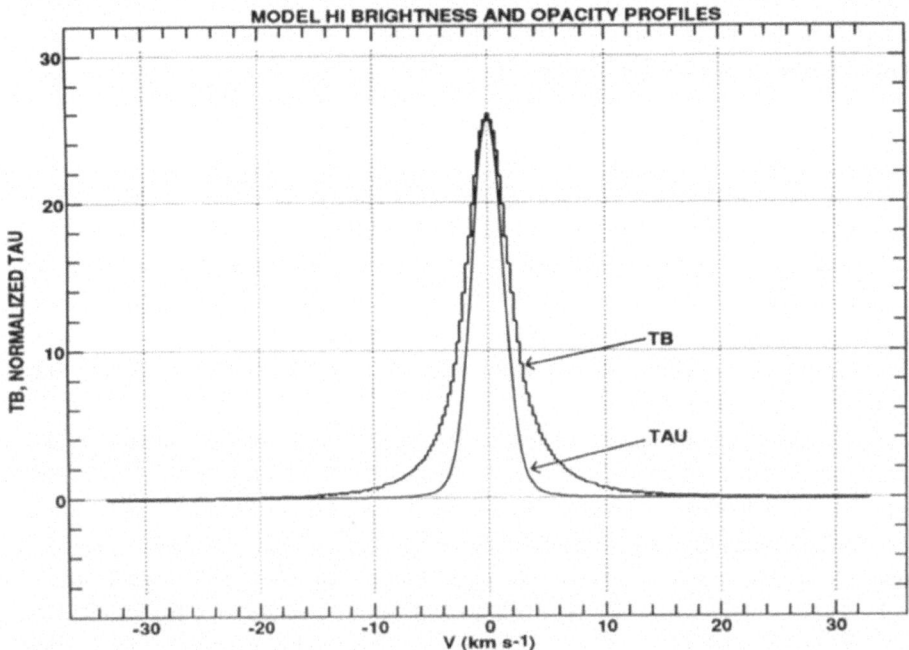

Fig. 6. Brightness temperature and optical depth line profiles from the heuristic model (the latter is normalized; $\tau(0) = 0.223$, corresponding to $T'_{sp} = T_b(v)/(1 - e^{-\tau(v)}) = 125K$). The profile of $T_b(v)$ has three significant gaussian components of width 3.4, 7.0, and 17.2 km s$^{-1}$, the opacity has two of width 3.1 and 7.2 km s$^{-1}$.

further and to show the effect of such structure on *optical/uv* absorption line profiles.

A moment's reflection will predict much of the detailed discussion below. When clouds have the structure imposed on them by the HII observations, species like NaI, CI, and FeI which appear preferentially in denser, cooler regions will have velocity profiles which are narrower, like $\lambda$21cm HI observed in absorption; by contrast, the profiles of the first ions of such species will be somewhat broader, akin to HI observed in emission, simply because their abundance is large in both the hotter and colder gas. This is a physical effect, arising from the fact that the line profiles are blends, in different proportions according to the degree of segregation, of narrower and broader components. However, line strength effects arise as well. Even for CI, there will be some contribution to the line profile from the broader, hotter material and, if the central optical depth of a given line is large enough to provide a substantial opacity in the broad wings, the effective width will be greater than that of a less opaque line of the same species. Thus it is to be expected that CII will have a curve of growth which differs from that of CI, and that curves of growth generally may not be characterized by a single b-value (fitted velocity width) across their expanse.

## 4.5. MODELLING THE HEURISTIC HI CLOUD

It is easy to model the velocity profiles needed for a simulation of optical/uv absorption lines. I did the following; 1) calculated the solid-angle averaged extinction $< A_V >$ at many radii in a supposed spherical cloud having the conditions in Figure 5 along each radius; 2) solved the equilibrium equations determining the ionization state and fine-structure level populations; 3) summed the profile contributions in each shell to form a global cloud profile for any species.

With one slight exception, there is little in the actual calculations which merits discussion. The extinction is supposed small and could have been ignored, but I followed the prescription of Black and Dalgarno (1977), taking the photoionization rate of species x as $\Gamma_x = \Gamma_{x,a} \exp(- < A_V > \Gamma_{x,b})$. I solved the equations of statistical equilibrium for the $^3P_{0,1,2}$ levels of CI and $^2P_{1/2,3/2}$ levels of CII in each radial zone, accounting for excitation by electrons and hydrogen atom impact, and solved the coupled ionization equilibrium equations of C and H as discussed below. In each radial zone the abundances and intrinsic linewidths are fixed by the considerations already elaborated and it is a simple matter to form global profiles by adding, at each velocity, the contribution from each zone. I used a velocity interval of $0.26$ km s$^{-1}$ and formed profiles in 256 'channels' extending $\pm$ 33 km s$^{-1}$ about the line center. As a benchmark, line profiles of the NaI U-and D-lines were formed, accounting for the 1771MHz hyperfine splitting.

### 4.5.1. Ionization Equilibrium and Electron Density

The usual simplifying assumption that all electrons arise from carbon is a surprisingly bad one for the range of conditions encountered in the model. Instead, it is necessary to solve the coupled ionization equilibrium of hydrogen and carbon. Defining $\xi_x$ as the abundance of species x relative to hydrogen, $\zeta_H(= 2 \times 10^{-17}s^{-1})$ as the primary cosmic ray ionization rate of HI, and $\alpha_x$ as the recombination rate coefficient, the equations of ionization equilibrium may be expressed as

$$n_{CII} = \xi_C n_{HI}/(1 + n_e \alpha_C/\Gamma_C), \tag{4a}$$

$$n_{HII} = n_{HI}/(1 + n_e \alpha_H/\zeta_H), \tag{4b}$$

$$n_e - n_{CII} - n_{HII} = 0. \tag{4c}$$

where $\Gamma_{C,a} = 1.3 \times 10^{-10} s^{-1}$. Recombination rates for H and C are given in Morton (1975) and Spitzer (1978).

Equation 4c may be recast as a cubic whose solution is root $x_2$ of equation 5.5.9 of Press *et al.* (1986), implemented in the Turbo Pascal procedure below which returns $n_e$ and $n_{CII}$ for given $n_{HI}$ and $T_k$. For undepleted carbon ($\xi_C = 3.75 \times 10^{-4}$) electrons from hydrogen dominate whenever $T_k > 10K$, $n_{HI} < 16 cm^{-3}$.

### 4.5.2. A little computer code

```
Procedure ElectronDensity(nH,T: double; var nE,nCII: double);
{solves coupled ionization equilibrium for C and H}
const
    zetaH: double = 2e-17;
    gammaC: double = 1.3e-10;
    xiC: double = 3.75e-04;
    alphaH: double = 8.01e-11;
    alphaC: double = 2.1e-10;
var
    betaH,betaC,Q,R,a1,a2: double;
begin
    T := exp(0.7*ln(T));
    betaH := zetaH*T/alphaH;
    betaC := gammaC*T/alphaC;
    a1 := (betaH+betaC)/3;
    a2 := betaC*betaH-nH*(xiC*betaC+betaH);
    Q := sqr(a1)-a2/3;
    R := sqr(a1)*a1-0.5*(a1*a2+betaC*betaH*nH*(1+xiC));
    nE := -2*sqrt(Q)*cos ((arctan(sqrt(Q*sqrt(Q/R)-1))+2*pi)/3)-a1;
    nCII := xiC*nH/(nE/betaC+1)
end;{ElectronDensity}
```

## 4.6. OPTICAL/UV ABSORPTION LINE PROFILES AND CURVES OF GROWTH

The species observed in absorption divide naturally into two groups, those which range widely across regions having different physical conditions (CII, FeII, OI) and those (CI,FeI, NaI) whose relative abundances increase sharply in the denser, cooler cloud core. In the model these spatial differences are manifested in velocity space as well, and illustrative examples are given in Figure 7 where I show the normalized velocity profiles dN/dV (column density per unit velocity) for CII $^2P_{1/2}$ and CI $^3P_0$. As suggested earlier, they exhibit the same disparities seen in HI emission and absorption at radiofrequencies. The CII line is fit by three gaussians with FWHM 3.1,

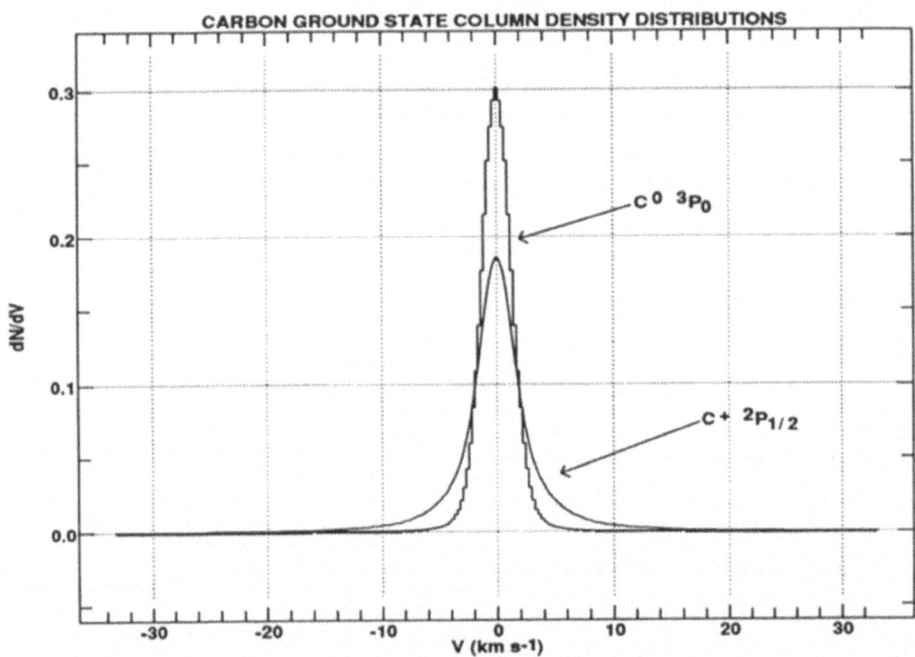

Fig. 7. Velocity profiles for CI and CII in the ground fine-structure state, normalized to unit area. Note the similarity between the CII profile and that of the $\lambda$21cm brightness temperature in Figure 6, and between CI and the profile of HI opacity.

6.6, and 16.6 km s$^{-1}$ having areas in the ratio 11:10:6; neutral carbon has two components with widths 2.9 and 6.3 km s$^{-1}$ and areas in the ratio 6.5:1. In both species, successively higher-lying fine-structure levels are more concentrated in the narrow core component.

The unusual velocity profiles arising in the heuristic HI model produce two related effects when curves of growth are calculated (Figure 8). For species like CII, there is slight tendency for the effective b-value to increase with increasing opacity; as $Nf\lambda$ increases, the CII equivalent width corresponds to somewhat different single-gaussian curves. An even stronger effect is that the curves of growth for NaI and CI lie markedly below that of CII for moderate opacities, but approach it in the limit of large optical depth; there is no flat portion to the curves of growth for the trace neutral species and their shape is somewhat bizarre. Note that the single-gaussian curves in these Figures extend to larger values of $Nf\lambda$ than actually obtain in the cloud model, as may be seen from the positions marked for the model NaI equivalent widths.

This simple model of an HI line of sight is grounded in qualities of the observations which are *universal* and might, therefore, be expected to be

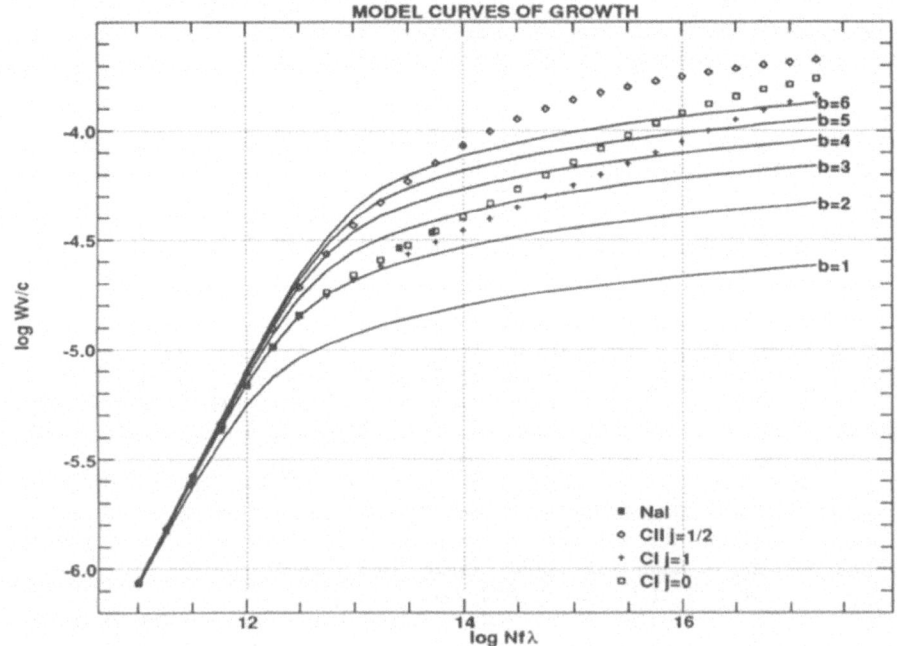

Fig. 8. Model cloud curves of growth ($W_v/c = W_\lambda/\lambda$) for species having column density distributions as shown in Figure 7 for carbon; also shown are the locations of the U-and D-lines of NaI in the model. The solid lines are classical curves of growth for several indicated b-values ($b_v = \sqrt{2}\sigma_v$).

manifest in optical absorption analysis. Unfortunately, such ordinary clouds as that modelled here are seldom isolated or interesting observationally. For the best-studied line of sight, toward $\zeta$Oph (Morton 1975), CI and NaI demonstrably lie on a curve of growth which is well below that of CII (b= 0.9 km s$^{-1}$ vs. 6.5 km s$^{-1}$) and has no flat portion. However, it is also true that the column density is large toward that star ($A_V \approx 1$), half the H-nuclei are in H$_2$, and $N_{CO} \approx N_{CI}$.

## Acknowledgements

The work described in Section 2 was done in collaboration with W. B. Burton. Valuable comments were provided by F. J. Lockman.

## References

Alvarez, H., May, J. & Bronfman, L.: 1990, *Astrophys. J.*, **348**, 495.
Baker, P., & Burton, W. B.: 1975, *Astrophys. J.*, **198**, 281.
Binney, J., Gerhard, O., & Stark, A.: 1991, *Mon. Not. R. Astron. Soc.*, **252**, 210.
Black, J. M. & Dalgarno, A.: 1977, *Astrophys. J. Suppl.*, **34**, 405.

Blitz, L. & Spergel, D. N.: 1991, *Astrophys. J.*, **370**, 205.
Burton, W. B.: 1988, in G. L. Verschuur and K. I. Kellermann, ed(s)., *Galactic and Extragalactic Radio Astronomy*, Springer, New York, 480.
Burton, W. B. & Gordon, M. A.: 1978, *Astron. Astrophys.*, **63**, 7.
Burton, W. B. & Liszt, H. S.: 1978, *Astrophys. J.*, **225**, 815.
Cassertano, S. & van Gorkom, J. H.: 1991, *Astron. J.*, **101**, 1234.
Ciardullo, R., Rubin, V., Jacoby, G., Ford, H. & Ford, W. K: 1988, *Astron. J.*, **95**, 438.
Clemens, D. P.: 1985, *Astrophys. J.*, **295**, 422.
Cohen, R. J. & Davies, R. D.: 1976, *Mon. Not. R. Astron. Soc.*, **175**, 1.
Colgan, S., Salpeter, E. E. & Terzian, Y.: 1986, *Astrophys. J.*, **328**, 275.
Dickey, J. M. & Lockman, F. J.: 1990, *Ann. Rev. Astron. Astrophys.*, **28**, 215.
Frail, D. A., Cordes, J. M., Hankins, T. H. & Weisberg, J. M: 1991, *Astrophys. J.*, , in press.
Fukui, Y., and Churchwell, E.: 1987, in D. C. Backer, ed(s)., *Conference Proceedings 155: The Galactic Center*, AIP, New York, 110.
Goldsmith, D. W., Habing, H. J. & Field, G. B.: 1969, *Astrophys. J.*, **158**, 173.
Goss, M., van Gorkom, J. H., Roberts, D. A. & Leahy, J. P.: 1990, in M. A. Gordon & R. L. Sorochenko, ed(s)., *Radio Recombination Lines: 25 Years of Investigation*, Dordrecht, Kluwer, 249.
Heeschen, D. S.: 1954, *An Investigation of the 21-cm. Line of Neutral Interstellar Hydrogen in the Section of the Galactic Center*, PhD Thesis, Harvard University.
Heiles, C. & Kulkarni, S.: 1988, in G. L. Verschuur & K. I. Kellermann, ed(s)., *Galactic and Extragalactic Radio Astronomy*, Springer, New York, 480.
Heiligman, G. M: 1987, *Astrophys. J.*, **314**, 747.
Kulkarni, S. & Heiles, C.: 1987, in D. J. Hollenbach, and H. A. Thronson, ed(s)., *Interstellar Processes*, Dordrecht, Reidel, 87.
Liszt, H. S.: 1983, *Astrophys. J.*, **275**, 163.
Liszt, H. S.: 1992, in J. P. Bergeron, ed(s)., *Highlights of Astronomy, vol. 8*, Dordrecht, Kluwer, in press.
Liszt, H. S. & Burton, W. B.: 1978, *Astrophys. J.*, **226**, 790.
Liszt, H. S. & Burton, W. B.: 1980, *Astrophys. J.*, **236**, 779.
Liszt, H. S., Burton, W. B. & Xiang, D.-L.: 1984, *Astron. Astrophys.*, **140**, 303.
Liszt, H. S., Burton, W. B. & van der Hulst, J. M.: 1985, *Astron. Astrophys.*, **142**, 237.
Lockman, F. J.: 1984, *Astrophys. J.*, **283**, 90.
Lockman, F. J, Jahoda, K. & McCammon, D.: 1986, *Astrophys. J.*, **302**, 432.
McKee, C. F. & Ostriker, J. P.: 1977, *Astrophys. J.*, **218**, 148.
Morton, D. C.: 1975, *Astrophys. J.*, **197**, 85.
Munch, G.: 1952, *Astrophys. J.*, **116**, 575.
Press, W. H., Flannery, B. P., Teukolsky, S. A. & Vetterling, W. T.: 1986, *Numerical Recipes*, Cambridge, Cambridge, 146.
Radhakrishnan, V., Murray, J. D., Lockhart, P. & Whittle, R. P. J.: 1972, *Astrophys. J. Suppl.*, **24**, 15.
Rubin, V. C.: 1988, in W. L. H. Shuter, ed(s)., *Kinematics, Structure, and Dynamics of the Milky Way*, Dordrecht, Reidel, 379.
Spitzer, L.: 1978, *Physical processes in the interstellar medium*, Wiley, New York, 154.
Wevers, B. M. H. R.: 1984, *A Study of Spiral Galaxies*, PhD Thesis, Groningen.

# MOLECULAR CLOUDS AND YOUNG MASSIVE STARS IN
# THE GALACTIC DISK

LEONARDO BRONFMAN

*Departamento de Astronomía, Universidad de Chile*

**Abstract.** The distribution of molecular clouds in the Galactic disk is reviewed, including all the data available to date from the Columbia CO Galactic surveys[1]. Using the CS ($J = 2 \rightarrow 1$) transition as a tracer of dense molecular gas associated with embedded compact H II regions, we present and discuss the distribution of young massive stars in the third and fourth Galactic quadrants and examine the process of molecular cloud-cloud collisions as a mechanism for massive star formation.

## 1. The Overall Distribution of Molecular Clouds

The discovery of giant molecular clouds, in the 1970s, represents one of the greatest advances in the study of the interstellar medium. An important fraction of the interstellar gas, probably about half, appears to be in molecular form; the individual masses of molecular clouds are estimated to be as high as $5 \times 10^5$ to $5 \times 10^6$ $M_\odot$, and their diameters are of the order of 50 to 100 pc; a total of about 500 molecular clouds in that mass range may exist within the solar circle. Such giant objects, each more massive than an average globular cluster, are the sites of most of the current star formation in the Galaxy. Within the last few years, more than 70 molecules have been discovered in the interstellar medium, from simple diatomic ones ($H_2$, CO, and CN) to long linear carbon chains ($HC_{11}N$) and ring molecules ($C_3H_2$). The study of these molecules, some not in existence under terrestrial conditions, has opened a whole new field of science, interstellar chemistry.

The most abundant molecule is by far $H_2$, but since it has no dipole moment it cannot be observed at radio frequencies. It is visible from satellites in ultraviolet absorption, but interstellar extinction limits observations to within about 2 kpc from the Earth. The second most common molecule, CO, is typically four to five orders of magnitude less abundant than $H_2$, but emission from its $J = 1 \rightarrow 0$ rotational transition, at a wavelength of 2.6 mm, is detectable in virtually all molecular clouds. Over the last decade, the 2.6 mm line from the CO molecule has proven to be the best available tracer of the molecular component of the interstellar medium and is now as important to Galactic structure studies as the 21 cm line of H I (Burton 1988; Combes 1991 and refs. therein). CO is not only an excellent qualitative tracer, but when averaged over a large section of a cloud, CO emission integrated in velocity seems to be proportional to the $H_2$ column density (Strong *et al.* 1988 and refs. therein); thus CO is a very good quantitative probe of

[1] Based on observations collected at the European Southern Observatory, La Silla, Chile

*L. Blitz (ed.), The Center, Bulge, and Disk of the Milky Way, 131–154.*
© 1992 *Kluwer Academic Publishers.*

molecular gas. Although the 2.6 mm CO line is usually optically thick, because molecular clouds seem to be composed of small clumps which do not seriously overlap in space and velocity, CO emission can be considered in an average sense as effectively optically thin.

The most complete and homogeneus CO survey of the Galaxy available to date is the combined Columbia survey of the first and fourth Galactic quadrants, carried out with the northern 1.2 m Millimeter-wave Telescope, now at the Center for Astrophysics, and with its southern twin at Cerro Tololo, Chile (Cohen 1978; Cohen et al. 1986; Dame et al. 1986; Bronfman et al. 1988a). This survey covers a 2°5 strip in Galactic latitude in the first quadrant and a 4° strip in the fourth quadrant, with a spatial resolution of 0°.125 and a velocity resolution of 1.3 km s$^{-1}$ (Fig. 1), and has been recently extended to the second and third Galactic quadrants (Digel et al. 1991; May et al. 1992). Completion of the Columbia CO survey of the Galaxy in the southern sky with an instrument identical to the one used in the north avoided the difficult problems encountered in reconciling the classical northern and southern 21 cm surveys made with different beams, sensitivities, calibrations, etc. The worth of an instrument in the one-meter class is obvious: it can make a well sampled survey of the fourth quadrant of the distant Galactic plane in a few hundred days of observation, a task totally impractical for a 10 m instrument.

The Columbia survey sampled about 67% of the face-on Galaxy within the solar circle and allowed the derivation of the mean axisymmetric distribution of molecular clouds in the Milky Way, and of the total mass of molecular hydrogen in the Galactic disk within the solar circle (Bronfman et al. 1988a). Nearly all earlier derivations of the amount and distribution of molecular gas in the inner Galaxy had been based on CO surveys of the first Galactic quadrant alone (Cohen and Thaddeus 1977; Burton and Gordon 1978; Cohen et al. 1980; Thaddeus and Dame 1984; Solomon et al 1979; Solomon and Sanders 1980; Sanders et al 1984). Given the deviations from azimuthal symmetry detected by preliminary in-plane observations (Robinson et al. 1984; Cohen et al. 1985), extrapolation of the northern results to the south seems to be an extremely dubious procedure. As A. S. Eddington said in his Halley lecture of 1930, following Kapteyn, '...to try to understand the structure of the Galaxy through observations made only from the northern hemisphere is like a bird trying to fly with one wing.'

Molecular clouds in the inner Galaxy are concentrated into a ring 3-4 kpc wide (the "molecular annulus"), which peaks at a distance of 5 kpc (for R$_0$ = 10 kpc) from the Galactic center and contains most of the molecular gas within the solar circle excluding the Galactic center (Scoville and Solomon 1975; Burton et al. 1975; Cohen and Thaddeus 1977; Sanders et al. 1984; Dame et al 1986; Bronfman et al 1988a). The molecular annulus derived from the Southern data alone is considerably flatter and broader than that

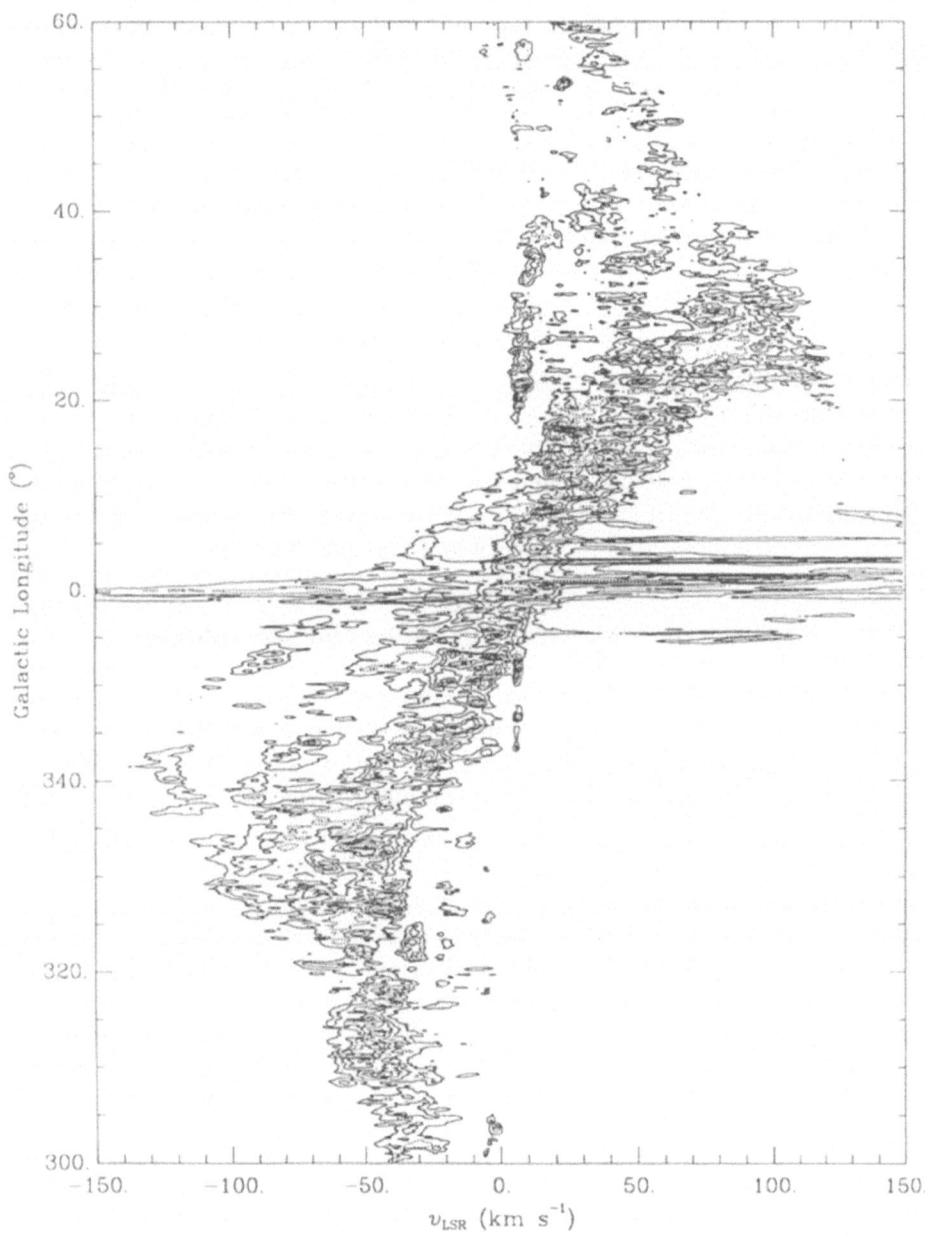

Fig. 1. Longitude-velocity diagram obtained by integrating the combined Columbia CO surveys of the first and fourth Galactic quadrants over latitude from -1° to 1° The resolution is 0°.125 by 1.3 km s$^{-1}$. The contour interval is 1 K deg.

derived from the Northern data alone (Fig. 2a). The mean half-width at half-maximum of the molecular layer, 70 pc in the inner Galaxy, is nearly the same in both hemispheres (Fig. 2b). The whole inner disk is gently warped in azimuth, the mean difference from north to south being about 17 pc. Beyond the solar circle there is a clear warp of the disk toward positive latitudes in the north and toward negative latitudes in the south.

The mass of molecular hydrogen between $R = 1.7$ kpc and $R = 8.5$ kpc has been reported to be about twice that of atomic hydrogen (Sanders et al. 1984; Rivolo and Solomon 1988), while Bronfman et al. (1988a) derived a rough equipartition between atomic and molecular gas within the solar circle. The latter authors have explained the difference as due to three factors with approximately similar weights; different instrumental calibrations, different conversion factors between CO integrated intensities and $H_2$ column densities, and different procedures to adjust the parameters of the axisymmetric model for a best fit to the observations; their analysis is shown to be self consistent in that it can reproduce the observed longitudinal distribution of CO intensity integrated in velocity and Galactic latitude.

## 2. Face-on View of Giant Molecular Complexes

Several procedures have been used to suppress the extended background emission in which giant molecular complexes in the inner Galaxy seem to be inmersed. Clipping at a fixed threshold, the method adopted by Myers et al. (1986), effectively distinguishes large complexes from the background, but it does not take into account the variations of the background emission expected with varying longitude and velocity. The approach we have adopted to identify the largest molecular complexes in the fourth Galactic quadrant, following Dame et al. (1986), is subtraction of a model background from the observed data set. It has been assumed in our model that the background emission is axisymmetric and has the same radial distribution as the overall CO emission shown above; that it is optically thin, and that it has a velocity dispersion of 8 km s$^{-1}$. At the adopted background level, 63% of the non-local ($|v| > 20$ km s$^{-1}$) observed emission has been removed, and the integrated intensities of the complexes determined directly from the resultant data set (Fig. 3). The kinematic distance ambiguity for molecular clouds within the solar circle has been removed through the help of associated optical objects, distance to the Galactic plane, associated radio H II regions, and $H_2CO$ absorption (Dame et al. 1986). The present analysis includes 80 molecular complexes between $l = 300°$ and $l = 348°$, 30 more than in Bronfman et al. (1989b). Molecular masses have been calculated using the $W(CO)/H_2$ conversion factor $X = 2.3 \times 10^{20}$ K km s$^{-1}$cm$^{-2}$ derived by Strong et al. (1988) from his analysis of gamma-ray, H I, and Columbia CO data. The mean molecular weight per $H_2$ molecule (including He) has been

a) Molecular gas mass surface density

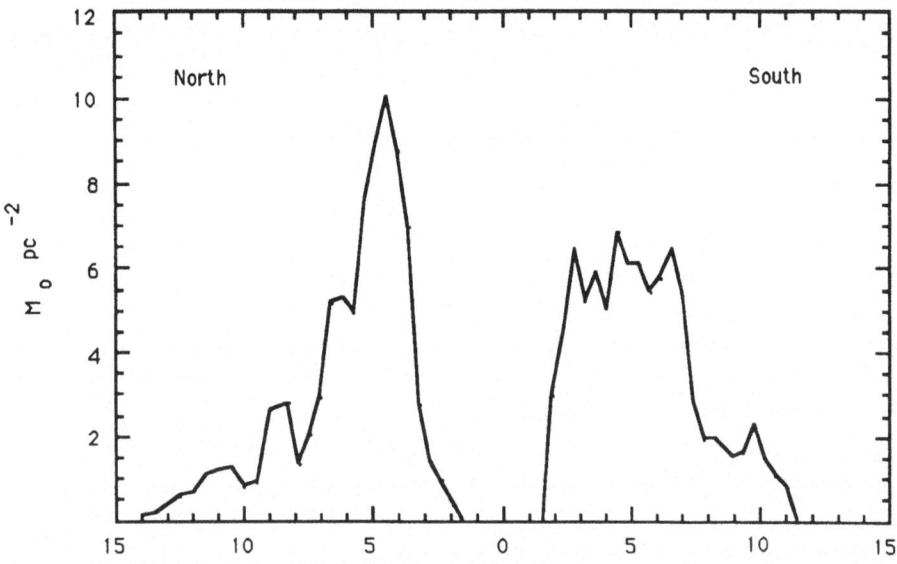

b) Centroid and half-width at half-maximum

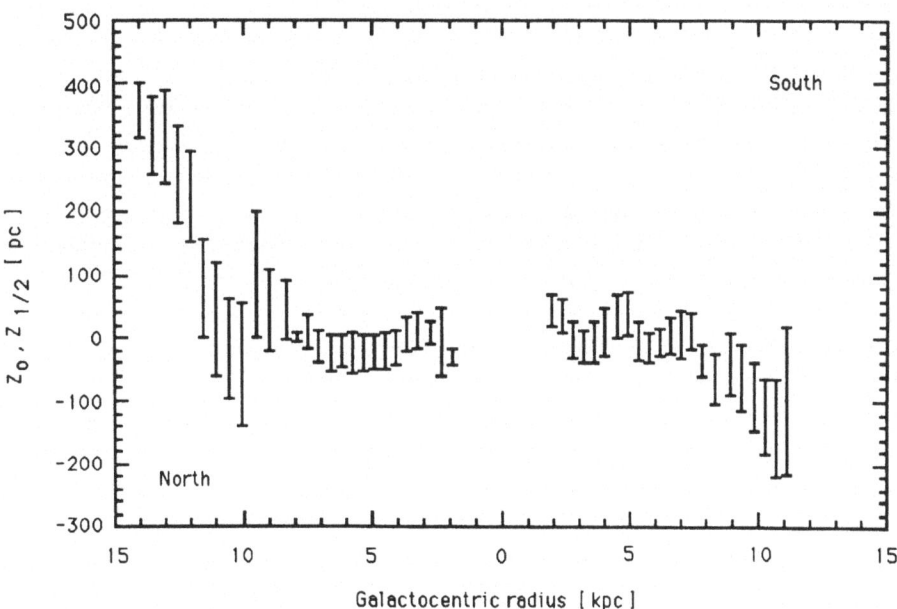

Galactocentric radius [ kpc ]

Fig. 2. a) Molecular gas mass surface density in the Galactic disk derived from the Columbia CO surveys of the first, second, and fourth Galactic quadrants. b) Centroid and half-width at half-maximum of the molecular disk as a function of Galactocentric radius.

taken to be $2.76m_H$.

The majority of the large scale surveys of carbon monoxide emission from the Galactic disk have been made using the most abundant $^{12}CO$ isotope; a relatively small number of $^{13}CO$ surveys have been published. Although weaker than $^{12}CO$ emission, because of its narrower linewidths and less saturated lines, the $^{13}CO$ emission facilitates separation of individual cloud contributions to the total emission of the Galactic plane. Liszt *et al.* (1984) find that the ratio of $^{12}CO$ to $^{13}CO$ emissivity, averaged over rings 0.5 kpc wide, decreases by more than a factor of 2 from the solar neighborhood to the peak of the molecular annulus, where the mean molecular gas density is much higher. They suggest that changes in the mean column density of molecular clouds over the Galactic disk dominates the observed variation of the emissivity ratio with R. There is a striking similarity, though, between the observed $^{13}CO$ $(l, v)$-diagram at b = 0° (Fig. 4; Bronfman *et al.* 1988b) and the corresponding $^{12}CO$ $(l, v)$-diagram with the model background subtracted (Fig. 3). If the emissivities are calculated for well defined molecular clouds, where $^{13}CO$ is readily detected, it is found that the emissivity ratio is rather constant ($\sim 5$) over the disk.

Maps of large areas of the Galactic plane made in the $^{12}CO$ and $^{13}CO$ isotopes have been used by Polk *et al.* (1988) to derive an average emissivity ratio of about 7, larger than the values found for GMCs. These authors suggest that a significant contribution to the large scale CO emission of the Galaxy is made by low optical depth molecular gas. The implication is that although in the method used to define GMCs we are arbitrarily subtracting about 60% of the total emission in the inner Galaxy, we are probably taking out only a small fraction of the total molecular gas mass. On a different scale a similar "pea-soup" model, a swarm of relatively high density clumps embedded in a more tenuous pervasive substrate of molecular gas, has been suggested by Blitz *et al.* (1986). We will show below that the molecular complexes defined here not only seem to carry a large fraction of the molecular gas mass, but they are also associated with most or all the massive star formation in the Galaxy.

In addition to the well defined Saggittarius-Carina spiral arm (Grabelsky *et al.* 1988) and to the 3-kpc expanding arm, visible in the $(l, v)$-diagram (Fig. 1) as the ridge of emission going from $l = 340°$, $v = -120$ km s$^{-1}$ to $l = 10°$, $v = 0$ the Galaxy (Fig. 5) seems to be organized into large scale features roughly coincident with the major spiral arms proposed by Georgelin and Georgelin (1976) based on the distribution of H II regions. Perturbations in the velocity field, either systematic (Burton 1971), or caused by energetic events (Nyman *et al.* 1987) may introduce large uncertainties in the derived kinematic distances, in particular near the terminal velocities. Studies of spiral structure in the inner Galaxy based on CO observations should be preceded by a careful examination of these systematic effects and

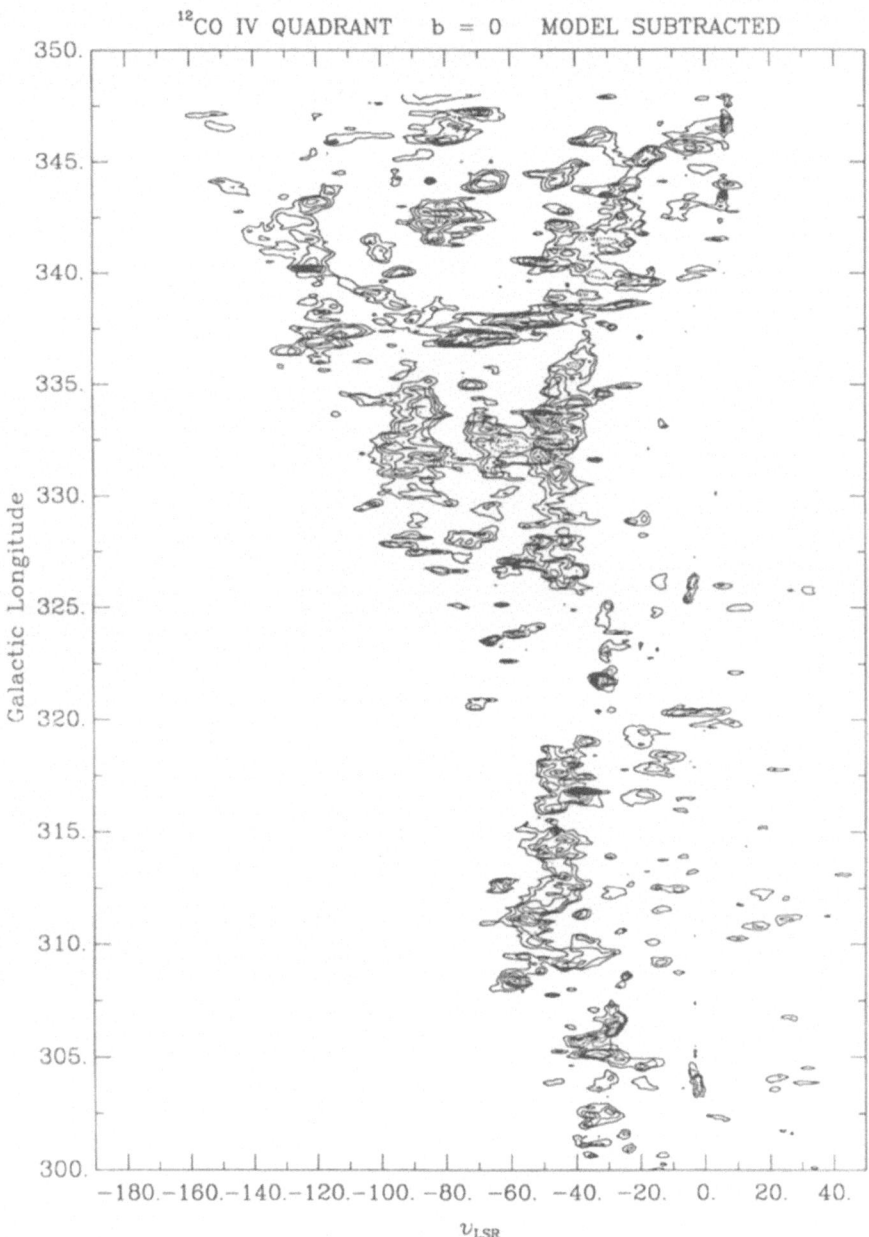

Fig. 3. Longitude-velocity diagram of $^{12}$CO emission at $b = 0°$ produced by subtracting our axisymmetric model background from the observed data in the fourth Galactic quadrant. The absolute scale of the background model was such that 63% of the emission in the observed data at $|v| > 20$ km s$^{-1}$ was removed. Contour interval is 0.8 K.

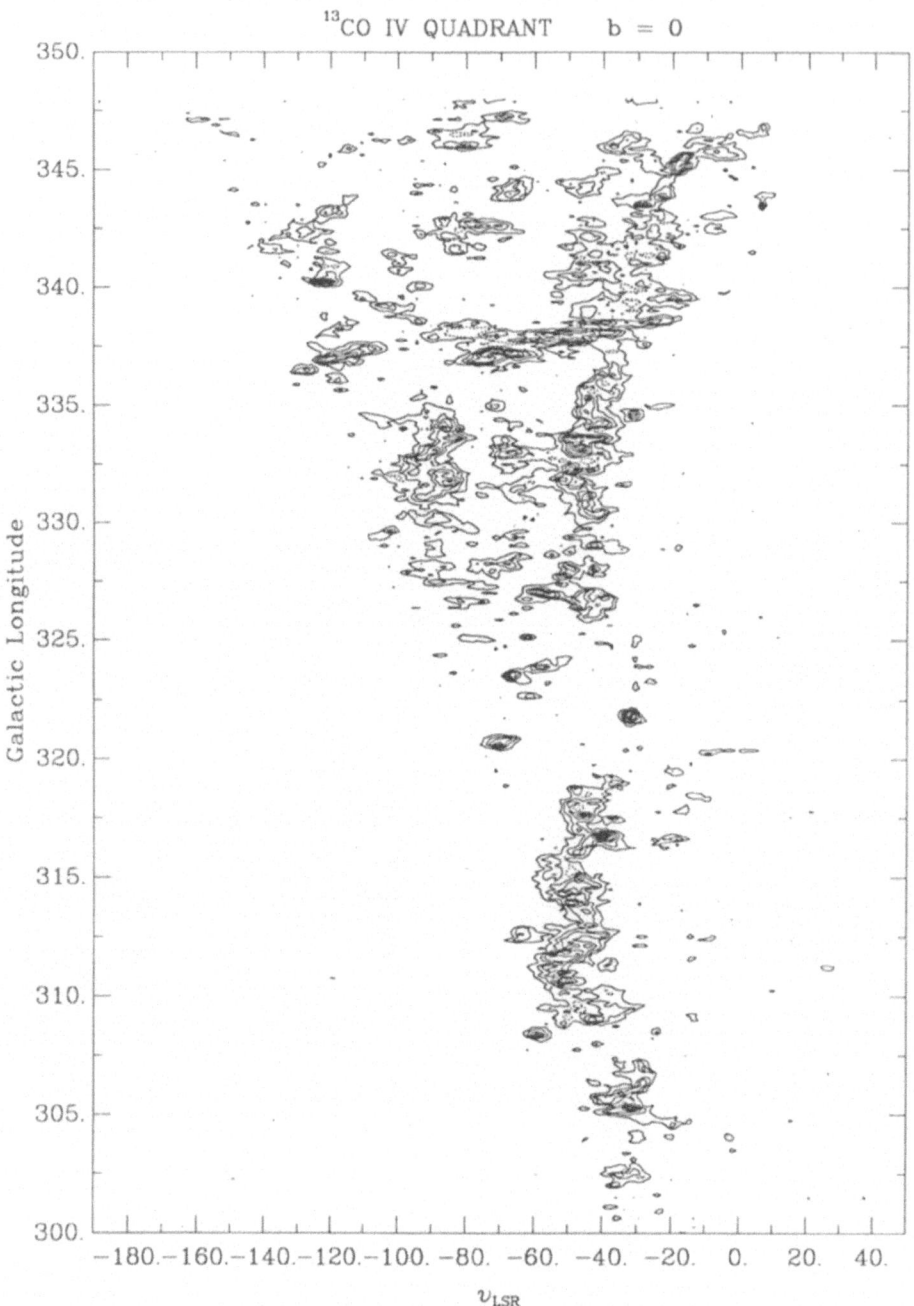

Fig. 4. Longitude-velocity diagram of $^{13}$CO emission of the region in (3). Contour interval is 0.3 K.

"FACE-ON" GALAXY

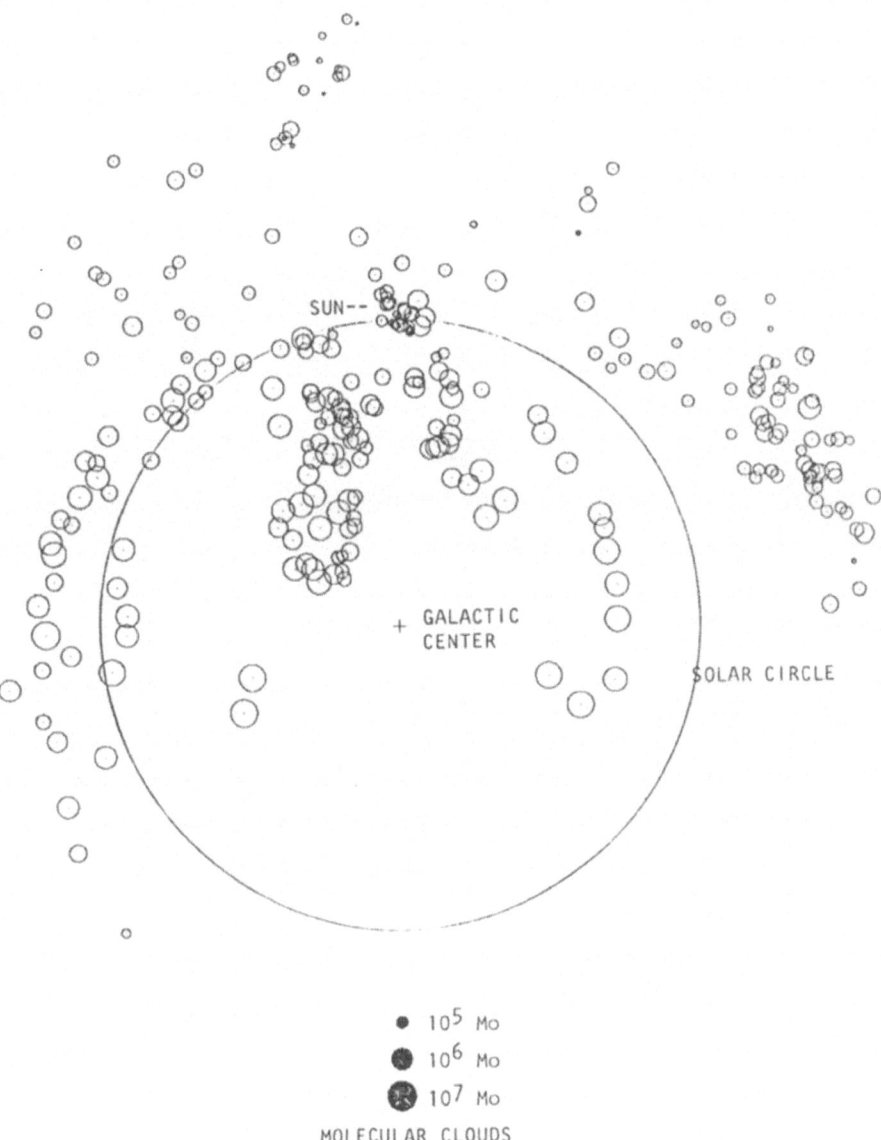

Fig. 5. Face-on view of molecular clouds in the Galaxy (open circles), including all the CO data obtained with the Columbia 1.2 m telescopes and analyzed to date. The size of the open circles in the figure is proportional to the logarithm of the molecular gas mass.

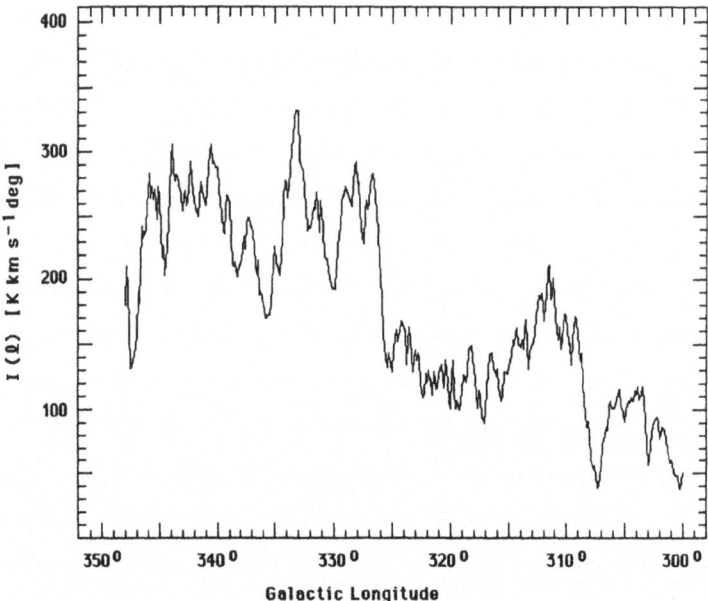

Fig. 6. Longitudinal distribution of CO emission integrated in velocity and latitude in the southern Galaxy. The steps at $l = 300°$ and $l = 328°$ may mark the tangent points of the Centaurus and Norma spiral arms, respectively.

of peculiar velocities associated with the molecular complexes. The best evidence, independent of molecular gas kinematics, for the existence of spiral arms in the inner Galaxy are the strong peaks in the run of CO emission integrated in velocity and latitude with Galactic longitude (Fig. 6).

## 3. Far Infrared Emission from the Galactic disk.

According to recent large scale comparison of IRAS, H I, CO, and radio continuum data, most of the FIR luminosity of the Galactic disk appears to be emitted by cold dust associated with diffuse H I (70%) and with molecular gas (20%) (Sodroski et al. 1989; Bloemen et al. 1990). Dust associated with H I is heated largely by the interstellar radiation field (ISRF), which may increase by a factor of 5 - 8 from the solar neighborhood to the inner Galaxy (Mathis et al. 1983), and is possibly dominated by low and intermediate mass stars (Bloemen et al. 1990). These authors suggest that the dust associated with molecular gas is also mostly heated by the ISRF. A different picture in which the molecular gas is heated by embedded and nearby OB stars is favored by Sodroski et al. (1989).

A large fraction of the giant molecular complexes in the Galaxy appear to be sites of active high-mass star formation. These luminous stars heat the surrounding dust which is readily detectable in the far infrared. Many of the complexes are associated with extended H II regions (Caswell and Haynes 1987; Fich and Blitz 1984) and with IRAS point-like sources, very likely to be associated with compact H II regions (Wood and Churchwell 1989; Wouterloot and Brand 1989; Bronfman et al. 1991). Even though on average the dust associated with the molecular component of the interstellar gas is colder than that associated with the atomic component, the infrared color temperature distribution obtained from the IRAS 60 $\mu$m and 100 $\mu$m bands has maxima at the site of giant molecular clouds in the Galactic plane (Burton et al. 1988).

What is the contribution of massive stars to the total FIR output of the Galactic disk? The contribution by warm dust associated with extended low density (ELD) H II regions ionized by massive stars has been estimated to be no more than 10% (Cox and Mezger 1989; Wood and Churchwell 1989). Within the solar circle up to 30% of the FIR emission may originate from OB star forming regions and ELD H II regions, according to Cox and Mezger (1989). Perault et al. (1989) suggest that OB stars may contribute about 70% of the total dust heating in the molecular annulus. Even though the FIR emission from the Galactic disk correlates better with H I than with $H_2$, there is a lack of FIR emission from the H I dominated outer Galaxy (Burton 1988) where there seems to be also a lack of massive stars (Conti et al. 1983). In the inner Galaxy, where most massive stars are found, the FIR emissivity per nucleon, like the ISRF, increases by about an order of magnitude from $R_\odot = 8.5$ kpc to $R = 3.5$ kpc (Bloemen et al. 1990; Perault et al. 1989). For normal spiral galaxies, massive stars have been claimed to be responsible for most of the FIR emission (Solomon and Sage 1988; Devereux and Young 1990). For a sample of 124 spirals, from types Sab to Sc, the latter authors have found that the ionizing stars required to explain the observed H$\alpha$ fluxes are enough to generate the observed FIR emission if their mass spectrum peaks at $\sim 16$ M$_\odot$. In a "leaky cloud" model (Boulanger and Perault 1988), the same stars which ionize the gas are heating the interstellar dust.

A still open question is whether the Galactic disk surface density of and the integrated luminosity associated with massive stars varies with Galactocentric radius. Boisse et al. (1981) found a defficiency of radio continuum emission compared to FIR emission in GMCs in the inner Galaxy, suggesting that there is a defficiency of very massive stars in the "molecular annulus" compared to the solar neighborhood and thus that the initial mass function (IMF) is not uniform with Galactocentric radius. Such a variation, however, has not been found by Myers et al. (1986) nor by Sodroski et al. (1987) in similar analyses. Estimation of the total number and integrated luminosity of O stars in the Galaxy has been normally made under the assumption of a uniform IMF (Wood and Churchwell 1989). To understand better the effects of massive star formation over the interstellar medium it appears useful to analyze the distribution and dominant mechanisms for massive-star formation in the Galaxy.

Using pre-IRAS far infrared data, radio continuum, the H $110\alpha$ recombination line, and the Columbia CO Survey of the first Galactic quadrant (Dame *et al.* 1986), Myers *et al.* (1986) found that most of the strong FIR sources in the Galaxy are associated with H II regions, and nearly all H II regions are in turn associated with molecular clouds. The distribution of massive molecular clouds in the Galactic plane is consistent with that of H II regions, particularly along the Sagittarius spiral arm. Clouds having no H II regions tend to have lower mass than clouds with H II regions, although there are massive molecular clouds ($M \geq 10^6$ $M_\odot$) having very few O stars. A cloud mass greater than $10^5$ $M_\odot$ appears to be a necessary but not sufficient condition for massive star formation. Inner Galaxy clouds, in particular, vary widely in their production of luminous stars.

For a sample of 94 giant molecular complexes associated with high luminosity H II regions, with masses in the range $10^5$ to $3 \times 10^6$ $M_\odot$, the efficiency (per unit mass of $H_2$) for OB star formation, as measured from Lyman continuum luminosity and H II region counting, decreases significantly with increasing cloud mass (Scoville *et al.* 1986). Massive star formation, according to these authors, is generally not stimulated by internal mechanisms such as sequential star formation, which should be more efficient for larger clouds. A possible mechanism for massive star formation, suggested by the observed quadratic dependence of the Galactic density distribution of H II regions on the local density of $H_2$, is that OB stars form as a result of cloud-cloud collisions (Scoville *et al.* 1986). The preference of OB star formation in spiral arms is then explained by the increased collision frequency of molecular clouds there. Mooney and Solomon (1988) find, however, that the star formation efficiency estimated from the ratio of FIR luminosity-to-cloud virial mass, for a sample of 55 GMCs associated with the most luminous H II regions in the first Galactic quadrant, is independent of cloud mass and varies widely over two orders of magnitude. This has been used as evidence against star formation initiated by cloud-cloud collisions, which would favor smaller clouds with larger collision cross-section per unit mass.

In the above discussion it has been implicitly assumed that the mass of a molecular cloud is proportional to its volume. However, because the virial mass of a molecular cloud is proportional to R $\Delta v^2$, the empirical size-linewidth relation R $\alpha$ $\Delta v^2$, valid for molecular clouds over six orders of magnitude, implies that the mass for a virialized cloud is proportional to its area. There is no guarantee, however, that molecular clouds are virialized. The relation between virial mass and CO luminosity obtained by Solomon *et al.* (1987) is a consequence of the size-linewitdh relation for clouds with the same average (over area and velocity) radiation temperature (Maloney 1990). The key point is that the collisional cross section per unit mass for virialized molecular clouds could be independent of cloud mass.

The CO and FIR emission has been compared by Scoville and Good (1989) for a sample of 41 clouds containing over half the giant radio H II regions in the first Galactic quadrant and for a comparison sample consisting of 7 clouds without H II regions. They found that massive star-forming clouds have higher peak FIR color temperatures, but when averaged over the cloud area, their FIR color temperatures (32 K) are not much different from the mean disk temperature (29 K). According to the measured FIR excesses for the H II region GMCs, only about 25% of the FIR luminosity originates from ionizing stars. Because the H II region GMCs have larger FIR to CO luminosity ratios than the average GMCs, they have a higher overall level of star formation activity. Therefore, H II region GMCs have larger rates of both high- and low-mass star formation. Their star-formation efficiency, as measured by Scoville and Good (1989) from the FIR to CO luminosity ratio, is independent of cloud size or mass.

These results have been extended to the outer Galaxy by Carpenter *et al.* (1990) using a sample of 21 molecular clouds associated with bright point-like sources in the second Galactic quadrant. The ratio of the FIR luminosity to cloud mass for their sample of clouds does not depend on cloud mass, and has similar values to those found in larger clouds in the inner Galaxy. Their interpretation is that there is no systematic difference between the rate of massive star formation in the inner and outer regions of the Galaxy. Because inner Galactic clouds lie in higher cloud density regions than outer Galactic clouds in their study, the implication is that if cloud-cloud collisions do indeed induce massive star formation, more collisions do not necessarily enhance the star formation rate per unit mass within a single cloud.

## 4. Massive Stars Embedded in Molecular Clouds

Most of the available studies of massive star formation in the Galaxy have been based on the comparison of large scale CO surveys and IRAS sky-flux images of the Galactic plane. Because, at least in the inner Galaxy, there are several molecular clouds at different distances in most lines of sight, it is not easy to estimate in that way the amount of infrared emission produced by a particular cloud and thus to study the star formation efficiencies. Our approach to study the Galactic distribution of young massive stars and the dominant mechanisms which stimulate massive star formation in the Galaxy is based on the determination, through the observation of millimeter-wave lines which are good tracers of high-density gas, of the FIR luminosity of dust associated with dense molecular cores and heated by embedded stellar sources.

Massive stars ($M \geq 10\ M_\odot$) in the Galaxy appear to be born in the dense cores of large molecular clouds (Bally 1989; Myers 1991). Dust in these cores is heated by ultraviolet light from the newly formed stars, reradiating their

energy mostly in the far-infrared, with a characteristic spectral signature
(Wood and Churchwell 1989). A good fraction of these young massive stars
seem to be found in clusters (Lada and Lada 1991); bright cluster members can be detected in the near infrared at 2.2 $\mu$m, and their membership
confirmed by JHK photometry (Lada et al. 1991). Millimeter wave emission lines from the associated molecular gas can be used to derive kinematic
distances for the dense cores, allowing in principle a determination of their
Galactic distribution and FIR luminosities. Wouterloot and Brand (1989)
have surveyed these dense regions in the outer Galaxy using the CO (1 → 0)
line (see also Carpenter et al. 1990). Toward the inner Galaxy, where most
of massive star formation takes place (Conti et al. 1983), the CO line is
affected by larger optical depths and, although being a good tracer of the
total molecular mass in a statistical sense, does not yield information about
the dense cores within molecular clouds. More specific observations of high
density tracers are needed in such case (see Zhou et al. 1991 and references
therein).

We have surveyed with the SEST (Swedish -ESO sub-millimeter wave
telescope) the CS (2 → 1) line, normally excited at $H_2$ densities in excess of
$10^4$ cm$^{-3}$, toward 800 point-like sources with FIR colors of embedded ultra-
compact H II regions in the third and fourth Galactic quadrants, detecting
420 of them with CS antenna temperatures larger than 0.2 K (Bronfman et
al. 1991; 1992). These sources had been previously identified from their FIR
colors to be probable sites of massive star formation (Wood and Churchwell
1989). There is a striking correlation between the FIR fluxes in the different
IRAS bands for all the point-like sources in which CS has been detected (Fig.
7). The FIR colors are very nearly the same for all these sources, so the temperature distribution of the dust surrounding the embedded stars appears to
depend only on the dust properties, which apparently do not change much
over the Galaxy. The temperature distribution appears to be unrelated to
the FIR luminosity of the source, which in this scheme would just measure
the total amount of illuminated dust, proportional to the energy output of
the heating source.

There is an excellent spatial correlation between massive star-forming
dense cores, as detected in the CS line, and velocity-integrated CO contour
maps obtained from the Columbia CO Survey of the Fourth Galactic quadrant (Bronfman et al. 1989a), strongly supporting the notion that OB stars
are formed in the dense cores ($N(H_2) \geq 10^4$ cm$^{-3}$ ) of large molecular clouds
with masses normally in excess of $5 \times 10^5$ M$_\odot$(Fig. 8a). An interesting finding in our survey is the presence of active high-mass star formation in the
3-kpc expanding arm (Fig 8b), a region previously characterized as having
little, if any, star formation. Extended mapping of several sources in the CS
line shows that massive-star forming cores have sizes between 1 pc and 3 pc,
and are largely similar in shape, mostly spherical or slightly elongated (Fig.
9).

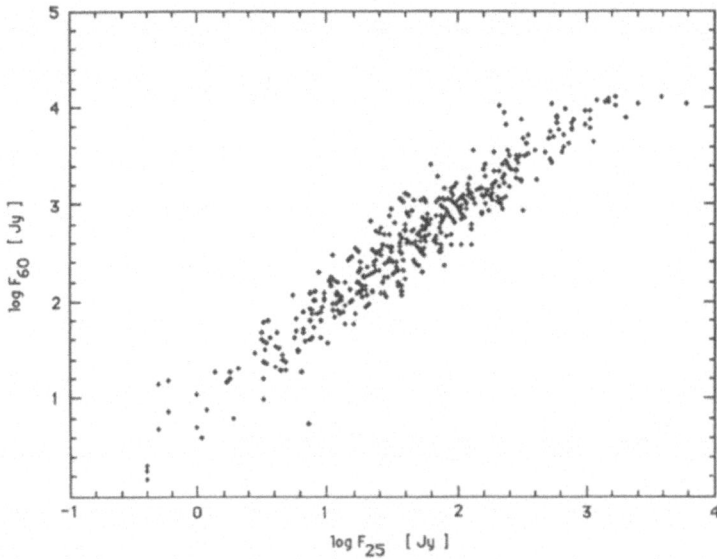

Fig. 7. Log of the 60 $\mu$m infrared flux against log of the 25 $\mu$m infrared flux for IRAS point-like sources associated with massive star forming regions for which the CS $(2 \rightarrow 1)$ line has been detected.

The kinematic information contained in the CS $(2 \rightarrow 1)$ line profiles allows, through the use of a standard rotation curve, a determination of Galactocentric distances for the detected sites of massive star formation. The SEST beamwidth of $45''$ at the CS $(2 \rightarrow 1)$ line frequency is comparable to the IRAS survey resolution; the reasonable correlation between FIR fluxes and CS line intensities (Fig. 10) suggests that we are probably sampling dust and gas within the same spatial regions. To derive the FIR luminosity of the dense cores, the twofold ambiguity in distance within the solar circle has been solved for each point-like source through its association with previously identified molecular clouds in the fourth Galactic quadrant. Using this method the FIR luminosity associated with embedded stellar sources has been computed for each molecular cloud identified.

Although all IRAS point-like sources detected in the CS $(2 \rightarrow 1)$ line are associated with molecular clouds, not every molecular cloud shows signs of massive star formation. For quiescent molecular clouds the dust heating must be mostly external and its main source the local ISRF. For those clouds that are undergoing massive star formation there appears to be a linear relation between FIR luminosity of embedded stars and the CO luminosity derived from our Columbia CO data (Fig. 11). It should be remembered that the CO

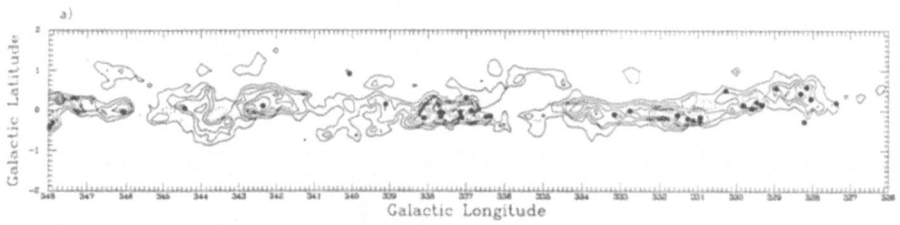

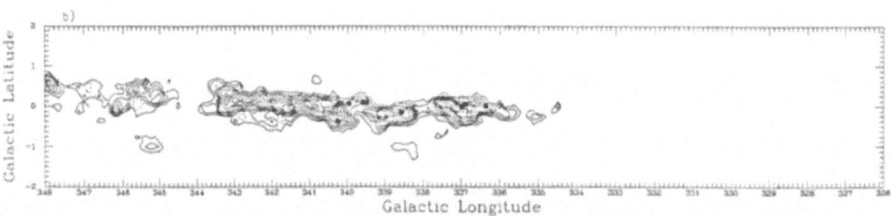

Fig. 8. Spatial maps of CO emission integrated in velocity. Embedded massive stars with their associated CS emission within the specified velocity ranges are plotted as filled circles. a) Velocity integration range goes from -100 km s$^{-1}$ to -60 km s$^{-1}$, and contour interval is 8 K km s$^{-1}$. b) Velocity integration range is a strip 15 km s$^{-1}$ wide about the position of the 3 kpc arm in the $(l, v)$ diagram. Contour interval is 3.5 K km s$^{-1}$.

luminosity is in turn proportional to the total FIR luminosity computed from IRAS sky-flux images for isolated molecular clouds associated with bright H II regions (Mooney and Solomon 1988). Embedded massive stars seem to be, according to our results, largely responsible for the heating of molecular clouds associated with H II regions in the Galactic plane. Alternatively, as pointed by Scoville and Good (1989), these clouds have larger rates of both high- and low-mass star formation, so the FIR luminosity associated with embedded massive stars may be tracing the overall star formation activity in the cloud.

The radial distribution of embedded massive stars in the southern Galaxy (Fig. 12), determined using the kinematic information contained in the CS line, rises smoothly from the 3 kpc arm to a maximum at R $\approx$ 5.5 kpc,

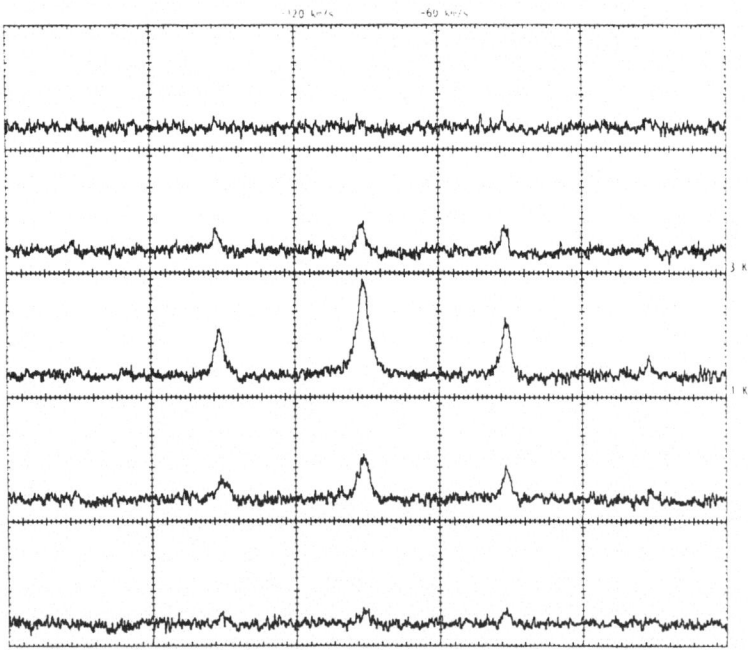

Fig. 9. Grid of CS $(2 \rightarrow 1)$ spectra, with a spacing of $45''$ centered at the position of IRAS16254-4300.

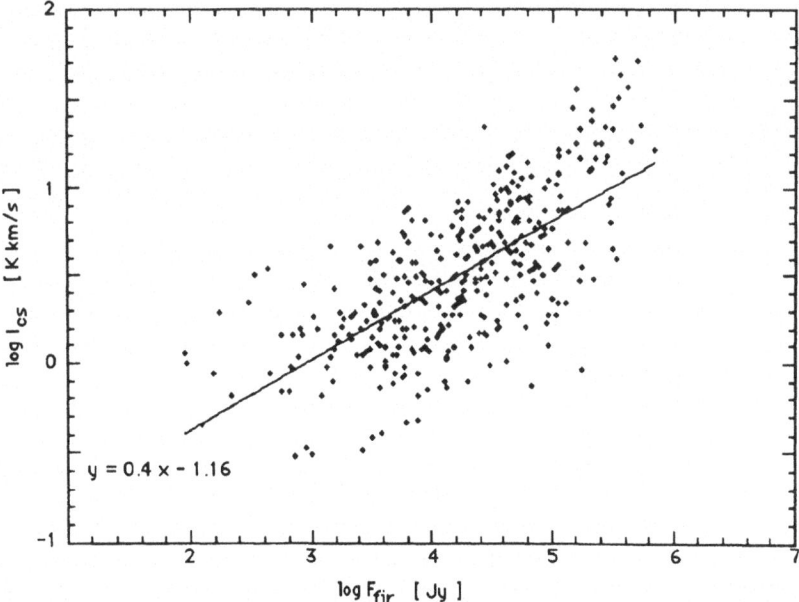

$y = 0.4 x - 1.16$

Fig. 10. Log of the CS emission integrated in velocity versus log of the equivalent IRAS flux (Boulanger and Perault 1988) for the massive star forming regions in our survey.

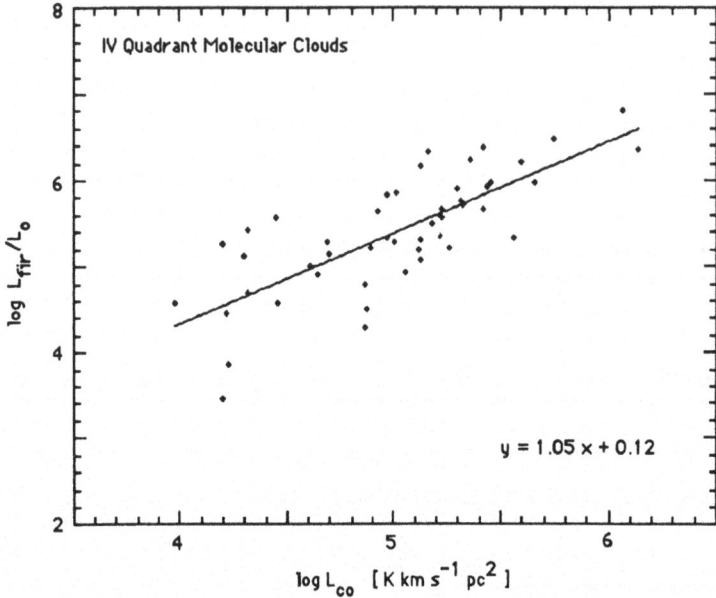

Fig. 11. Log of FIR luminosity associated with embedded stellar sources versus log of integrated CO luminosity for massive star forming molecular clouds in the fourth Galactic quadrant.

decaying exponentially toward the outer Galaxy with a scale length    of $\sim 2$ kpc, with massive star formation present out to $R = 19$ kpc. The distribution perpendicular to the Galactic plane (Fig. 13) is similar to that of molecular clouds, with a mean thickness of about 70 pc. In the outer Galaxy the distribution of embedded massive stars follows the well known warp toward negative latitude (May $et$ $al.$ 1985; 1988; Wouterloot $et$ $al.$ 1990). The average face-on FIR surface luminosity associated with embedded stellar objects has been computed for Galactocentric rings 1 kpc wide and compared to the $H_2$ surface density derived from our Columbia CO data (Fig. 14). The radial distribution of FIR surface luminosity associated with embedded massive stars resembles the molecular gas distribution, but the "massive-star" annulus, like the Galactocentric dependence of the Lyman continuum photon production rate (Gusten and Mezger 1983), is narrower    than the "molecular annulus". The dependence of massive star formation efficiency, measured here as the ratio of the FIR surface luminosity associated with embedded stars to the $H_2$ surface density, on Galactocentric radius (Fig. 15) is similar to that derived for the ISRF by Mathis $et$ $al$ (1983) and to the near infrared (2.4 $\mu$m) source function for the Galactic plane recently derived using IRT (Infrared Telescope) observations (Kent, Dame and Fazio 1991). It resembles also the FIR emissivity function derived for the Galactic

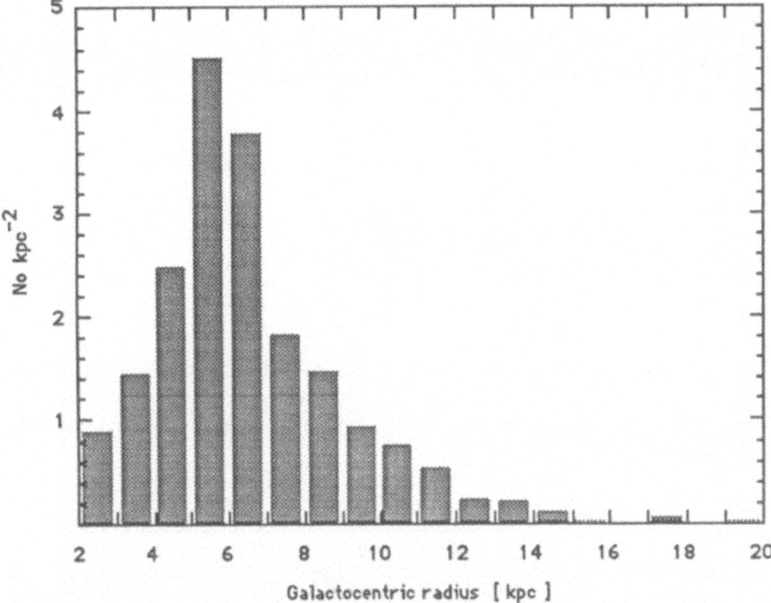

Fig. 12. Number of massive star forming regions per unit area binned in rings 1 kpc wide in the third and fourth Galactic quadrants.

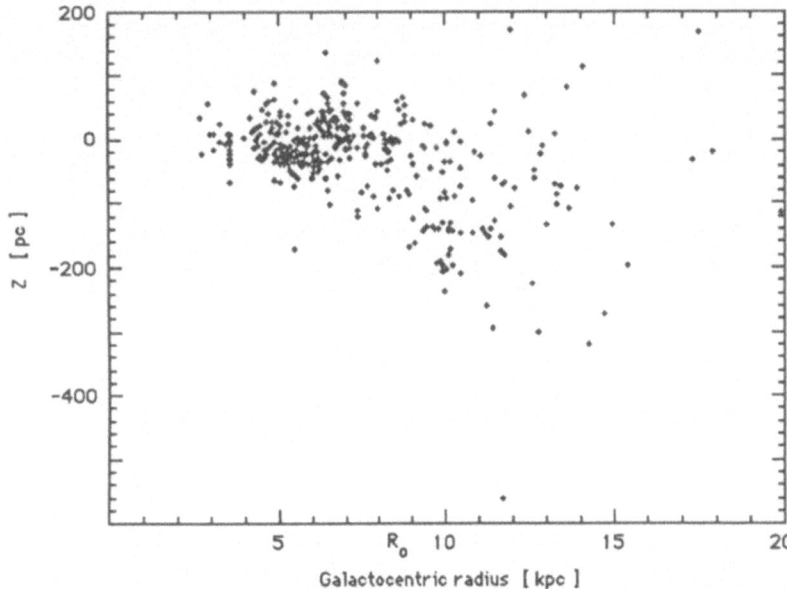

Fig. 13. Distance to the Galactic plane of embedded massive stars as a function of Galactocentric radius in the third and fourth Galactic quadrants.

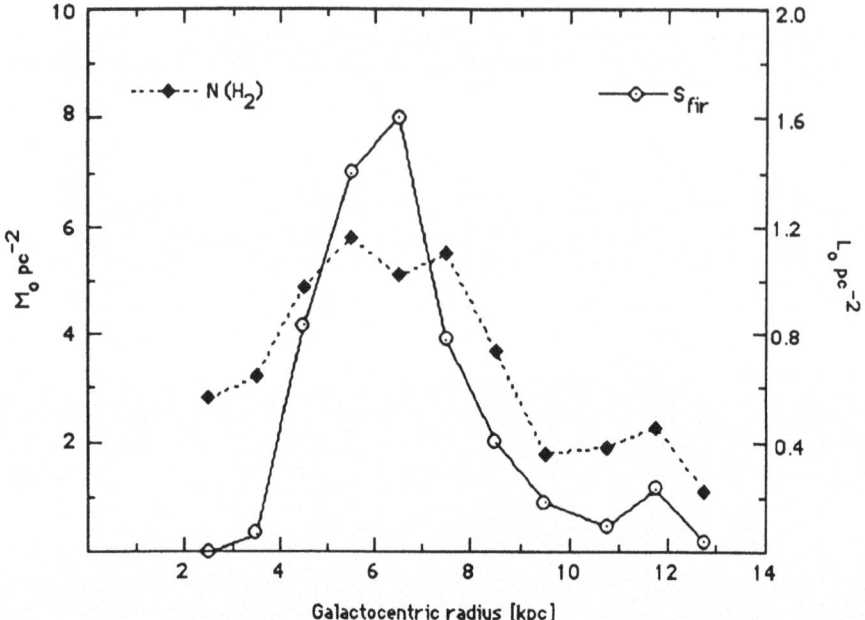

Fig. 14. Variation of the face-on FIR surface luminosity associated with embedded massive stars (open squares) and $H_2$ surface density (filled squares) with radius in the southern Galaxy from the 3-kpc arm to the Carina region.

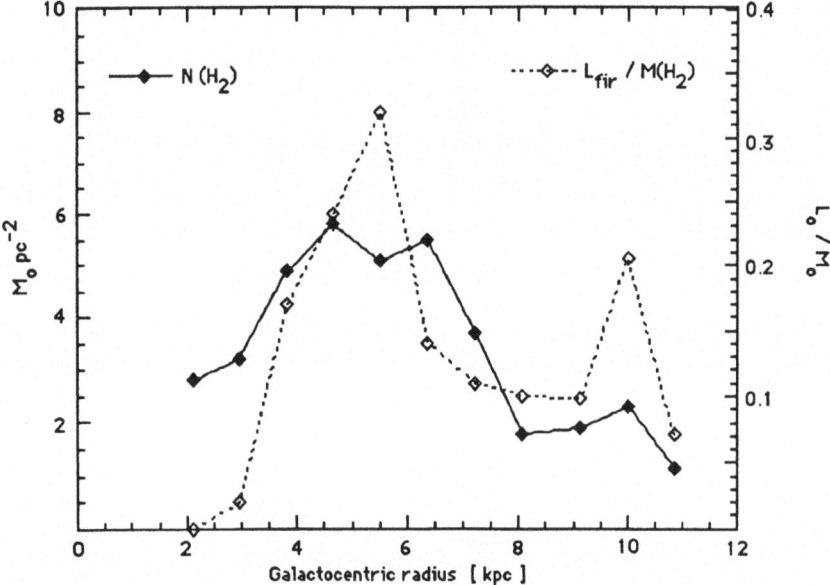

Fig. 15. Variation of the massive star formation efficiency (open squares) and $H_2$ surface density (filled squares) with Galactocentric radius. The region is the same as in (14).

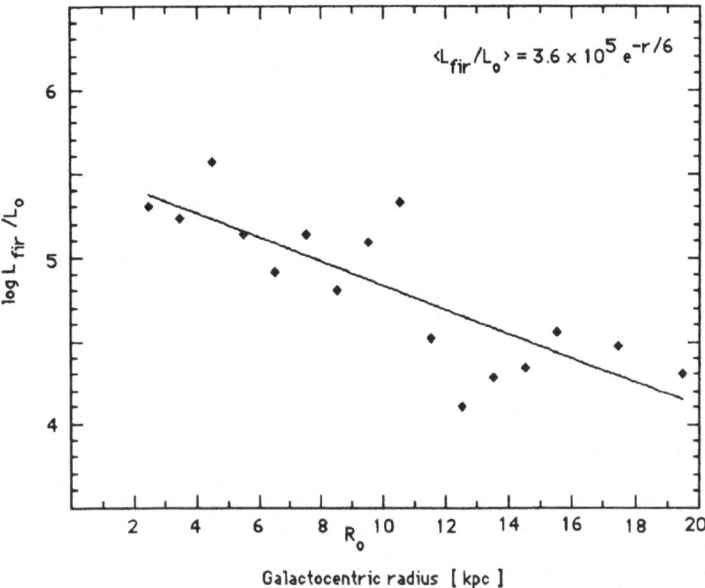

Fig. 16. Variation of the mean FIR luminosity of embedded massive star forming regions with Galactocentric radius. The region is the same as in (14).

plane by Bloemen *et al.* (1990).

It would appear that by studying the population of embedded massive stars we may be tracing the intensity of the local ISRF in the Galactic plane, largely responsible for the heating of dust associated both with atomic and cold molecular hydrogen. This could be the case if young massive stars with the same spatial distribution than our sample, but already free of their parental gas, were largely responsible for the ISRF. Alternatively, as pointed out above, the FIR luminosity of embedded massive stars may be tracing the overall level of star formation activity in the Galactic plane. Apparently there are not enough OB stars in the Galaxy to provide the dust heating (Cox and Mezger 1989). In this sense, because the average FIR luminosity of our sample of embedded stellar sources varies by more than one order of magnitude from the outer to the inner Galaxy (Fig. 16), it may be unsafe to extrapolate the IMF derived for the solar neighborhood to the rest of the Galactic plane.

Massive star formation in the Galactic disk, according to our results, appears to be stimulated largely by collisions between molecular clouds in the Galactic disk, as suggested by Scoville *et al.* (1986). For star forming molecular clouds the FIR luminosity associated with embedded massive stars is proportional to the cloud CO luminosity, and thus the star formation rate per unit $H_2$ mass is independent of cloud mass. However, if we compute

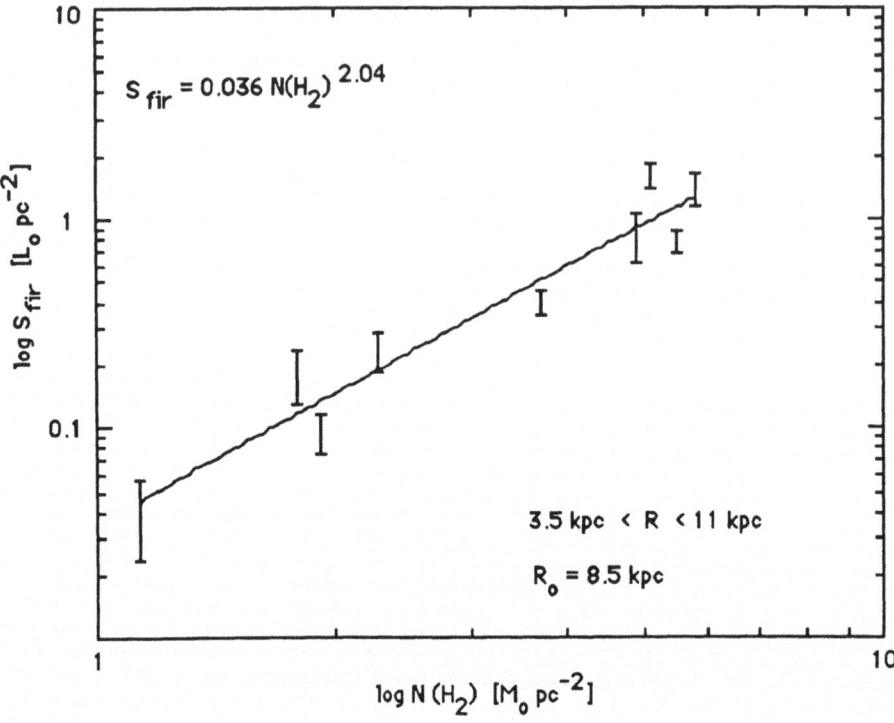

Fig. 17. Variation of the averaged face-on FIR luminosity with $H_2$ surface density computed for rings 1 kpc wide in the southern Galaxy from the 3-kpc arm to Carina.

the FIR luminosity over regions of the Galactic disk much larger than the mean free path between molecular clouds, and take into account both star-forming and quiescent molecular clouds in the gas budget, we find that for a large region of the Galaxy, from the 3 kpc expanding arm to the Carina region at R = 11 kpc, the face-on FIR surface luminosity associated with embedded massive stars in the Galactic disk is proportional to the $2^{nd}$ power of the $H_2$ surface density (Fig. 17). The massive star formation efficiency, measured as the ratio of the FIR surface luminosity associated with embedded stars to the $H_2$ surface density, seems to be proportional to the averaged $H_2$ surface density, as would be expected if cloud-cloud collisions were the dominant mechanism triggering massive star formation in molecular clouds in the Galaxy.

## Acknowledgements

I am indebted to Jorge May and Lars-Åke Nyman for use of data prior to publication, and to Patrick Thaddeus for his support during visits to the Harvard-Smithsonian Center for Astrophysics. Guido Garay Tom Dame, and Seth Digel provided useful comments and discussions. Felipe Manriquez gived assistance with the observations and helped editing the text. This work has been partially supported by Proyecto FONDECYT 90-1097, República de Chile and by NASA Grant NAG5-1179.

## References

Bally, J. 1989, in *ESO Conference and Workshop Proceedings* No. 33.

Blitz, L. and Stark, A. 1986. *Ap. J.* 300: L89.

Bloemen, J. B. G. M., Deul, E., Thaddeus, P. 1990, *Astron. Astrophys.* 233: 437.

Boisse, P., Gispert, R., Coron, N., Wijnbergen, J.J., Serra, G., Ryter, C., and Puget, J.L. 1981. *Astron. Astrophys.* 94: 265.

Boulanger, F., and Perault , M. 1988,*Ap. J.*330: 964.

Bronfman, L., Cohen, R., Alvarez, H., May, J., and Thaddeus, P. 1988a,*Ap. J.* 324: 248-266.

Bronfman, L., Bitran, M, Thaddeus, P. 1988b. *Lecture Notes in Physics* 315: 318.

Bronfman, L., Alvarez, H., May, J., Thaddeus, P. 1989a*Ap. J.* (Supp.) 71: 481.

Bronfman, L., Nyman, L., Thaddeus, P. 1989b. *Lecture Notes in Physics* 331:139.

Bronfman, L., May, J., and Nyman, L. 1991. *Proceedings of IAU Symposium No 147 "Fragmentation of Molecular Clouds and Star Formation."*

Bronfman, L., May, J., Nyman, L., and Thaddeus, P. 1992, in preparation.

Burton, W. B. 1971. *Astron. Astrophys.* 10: 76.

Burton, W. B., Gordon, M. A., Bania, T. M., Lockman, F. J. 1975,*Ap. J.* 202: 30.

Burton, W. B., and Gordon, M. A. 1978, *Astron. Astrophys.* 63: 7.

Burton, W. B. 1988, in *Galactic and Extragalactic Radio Astronomy*, ed. G. L. Verschuur and K. I. Kellerman (Berlin: Springer), p. 295

Carpenter, J., Snell, R., and Schloerb, F.P. 1990,*Ap. J.* 362: 147.

Caswell, J.L., and Haynes, R.F. 1987,*Astron. Astrophys.* 171: 261.

Cohen, R. S. 1978, Ph. D. thesis, Columbia University.

Cohen, R. S., Cong, H., Dame, T. , Thaddeus, P. 1980,*Ap. J.* (Letters) 217: L155

Cohen, R. S., Dame , T. M, and Thaddeus, P. 1986,*Ap. J.* Suppl. 60: 695.

Cohen, R. S., and Thaddeus, P. 1977,*Ap. J.* 217: L155.

Cohen, R. S., Thaddeus, P., Bronfman,L. 1985, *IAU Symp. 106, The Milky Way*, ed H. Van Woerden, R.J. Allen, W.B. Burton (Dordrecht: Reidel), p. 199.

Combes, F. 1991. *Ann. Rev. Astron. Astrophys.* 29: 195.

Conti , P., Garmany, C., de Loore, C., and Vanbeveren, D. 1983,*Ap. J.* 274: 302.

Cox, P., and Mezger, P. G. 1989,*Astron. Astrophys.* Rev.

Dame, T., Elmegreen, B., Cohen., R. S., and Thaddeus, P. 1986.*Ap. J.* 305: 892

Devereux, N., and Young, J. 1990,*Ap. J.* 350: L25.

Digel, S. 1991, Ph. D. Thesis, Harvard University.

Fich, M., and Blitz, L. 1984,*Ap. J.* 279: 125.

Georgelin, Y. M., and Georgelin, Y. P 1976, *Astron. Astrophys.* 49: 57.

Grabelsky, D., Cohen, R., Bronfman, L., Thaddeus, P. 1988.*Ap. J.* 331. 181.

Gusten, R., Mezger, P. G. 1983, *Vistas in Astronomy* 26:159

Kent, S. M., Dame, T., and Fazio, G. 1991, preprint.

Lada, E., DePoy, D., Evans II, N., and Gatley, I. 1991, preprint.

Lada, C., and Lada, E. 1991, preprint.

Liszt, H. S., Burton, W. B., and Xiang, D. L. 1984. *Astron. Astrophys.* 140: 303

Mathis, J. S., Mezger, P. G., and Panagia, N.1983,*Astron. Astrophys.* 128: 212.

May, J., Alvarez, H., Garay, G., Murphy, D., Cohen, R., and Thaddeus. P. 1985. *ESO-IRAM-Onsala Workshop on (Sub) Millimeter Astronomy.*

May, J., Murphy, D., and Thaddeus, P. 1988.*Astron. Astrophys.* Supp. 73: 51.

May, J., Alvarez, H., and Bronfman, L. 1992, in preparation.

Mooney, T. J., and Solomon, P. M. 1988,*Ap. J.* 334: L51.

Myers, P C., Dame, T. M, Thaddeus, P., Cohen, R.S., Silverberg, R.F., Dwek, E., and Hauser, M. G. 1986,*Ap. J.* 301: 398

Myers, P. C. 1991, preprint.

Nyman, L., Thaddeus, P., Bronfman, L., and Cohen, R. 1987.*Ap. J.* 314: 374.

Perault, M., Boulanger, F., Puget, J., and Falgarone 1989.*Ap. J.*

Polk, K., Knapp, G., Stark, A., and Wilson, R. 1988.*Ap. J.* 332: 432

Rivolo, A. and Solomon, P. 1988. *Lecture Notes in Physics* 315: 42

Robinson, B. J., et al. 1984,*Ap. J.* 283: L31.

Sanders, D. B., Solomon, P.M., and Scoville, N.Z. 1984,*Ap. J.* 276: 182.

Scoville, N. Z., and Solomon, P.M. 1975,*Ap. J.* 199: L105.

Scoville, N. Z., Sanders, D.B., and Clemens, D.P. 1986,*Ap. J.* 305: L45.

Scoville, N.Z., and Good, J.C. 1989,*Ap. J.* 339: 149

Sodrosky, T.J., Dwek, E., Hauser, M.G, and Kerr, F.J. 1987, in *Star Formation in Galaxies*, ed. C.J. Lonsdale Persson, p. 37.

Sodrosky, T.J., Dwek, E., Hauser, M.G, and Kerr, F.J. 1989,*Ap. J.* 336: 762.

Solomon, P. M., and Sanders, D. B. 1980, in *Giant Molecular Clouds in the Galaxy*, ed. P. M. Solomon and M. G. Edmunds (Oxford: Pergamon), p. 41.

Solomon, P. M., Scoville, N. Z., and Sanders, D. B. 1979,*Ap. J.* 232: L89.

Solomon, P. M., Rivolo, A.R., Barrett, J., and Yahil, A. 1987.*Ap. J.* 319: 730

Solomon, P. M., and Sage, L.J. 1988,*Ap. J.* 334: 613

Strong, A. W., Bloemen, J.B.G.M., Dame, T., Grenier, I.A., Hermsen, W., Lebrun, F Nyman, L., Pollok, A., Thaddeus, P. 1988. *Astron. Astrophys.* 207: 1.

Thaddeus, P. and Dame, T. M. 1984, *Proceedings of Workshop on Star Formation, in Occasional Reports of Royal Observatory, Edinburgh*, (ed. R. Wolstencroft).

Wood, D.O.S, and Churchwell, E. 1989, *Ap. J.* 340: 265.

Wouterloot, J. G. A., Brand, J.,1989,*Astron. Astrophys.* Suppl. 280: 149.

Wouterloot, J. G. A., Brand, J., Burton, W. B., and Kwee, K. K. 1990, *Astron. Astrophys.* 230: 21.

Zhou, S., Evans., N., Gusten, R., Mundy, L., and Kutner, M. 1991,*Ap. J.* 372: 518.

# THE DISTRIBUTION OF STARS IN THE DISK AND HALO
# OF THE GALAXY

DAVID W. LATHAM

*Harvard-Smithsonian Center for Astrophysics*
*60 Garden Street, Cambridge, Massachusetts 02138, U.S.A.*

**Abstract.** Globular clusters are a classical probe of the spatial distribution, kinematics, and metallicities in the halo of our Galaxy, but this approach is limited by the finite number of clusters. Surveys of nearby field stars with extreme kinematics or extreme metallicities can also contribute to the study of disk *versus* halo stars, but in this case the interpretation is difficult because of selection effects. Surveys of faint stars *in situ* at large distances show great promise for studies of the structure and evolution of our Milky Way Galaxy, because they can be designed to avoid most selection effects.

## 1. Introduction

The sun lies near the central plane of the disk of our Galaxy. Most of the stars in the solar neighborhood, about 90%, belong to the classical thin disk, with a scale height of approximately 250 pc. The Sun and its thin-disk neighbors move in nearly circular orbits around the center of our Galaxy, with a speed of about 225 $km s^{-1}$ . The thin disk is indeed quite thin; the distance to the center of the Galaxy is about 30 times larger than the local scale height.

A tiny fraction of the stars in the solar neighborhood, about 0.2%, belong to the halo, a very different population that has a roughly spheroidal distribution centered on our Galaxy, and hardly any net rotation. Three arguments support the view that the halo population formed as our Galaxy was in the process of collapsing from a giant cloud of gas and dust. First, the spatial distribution of the halo stars above and below the plane of the disk, and the distribution of their Galactic orbits, indicate that these stars were formed as the Galaxy was undergoing its initial collapse, before the disk could form by dissipative processes. The halo stars preserve a fossil record of the spatial distribution and kinematics of the material from which they formed, affected only by the changing potential of the collapsing Galaxy. Second, estimates of the ages of the halo stars place them among the oldest objects in the Galaxy. Third, the atmospheres of halo dwarf stars have preserved a fossil record of the chemical composition of the material from which they formed. Spectroscopy shows that the typical halo star is metal poor by a factor of 30 compared to the Sun, which supports the idea that the halo stars formed from material that had not yet been enriched very much by preceding generations of stars.

155

*L. Blitz (ed.), The Center, Bulge, and Disk of the Milky Way, 155–163.*
© 1992 *Kluwer Academic Publishers.*

About 10% of the stars in the solar neighborhood do not fit neatly into either the thin disk or the halo populations. These stars lie in a kind of thick disk distribution with a local scale height of perhaps 1,500 pc. Is the thick disk distribution just the tail of the thin disk, which has been puffed up by some heating mechanism in the disk? Or, did the thick disk originate in some separate episode early in the history of our Galaxy? Despite recent progress in figuring out the kinematics, spatial distribution, and metallicities of the disk and halo populations, the origin of the thick disk is still under intense debate (*e.g.* see Norris & Ryan 1991).

One of the most important parameters describing a star or a stellar population is its age. Unfortunately we do not yet have a reliable way to measure the ages of individual stars, especially older stars. The fact that stellar ages are hardly mentioned in this paper is a reflection of the uncertain status of age measurements.

In this paper I review briefly what is known about the disk and halo populations from studies of globular clusters, and from studies of field stars in the solar neighborhood. Then I present the results of an *in situ* study of thick disk and halo stars based on a survey of faint stars at the north Galactic pole carried out by Ken Croswell.

## 2. Globular Clusters

A classical probe of the Galaxy's halo is its family of globular clusters, which have long been known to be distributed above and below the plane of the Milky Way. In a seminal paper, Zinn (1985) showed conclusively that the globular clusters are actually divided into two families. The first family consists of clusters which are distributed above and below the disk more or less spheroidally, as expected for the halo, with a scale height of several thousand pc. Clusters in the second family have a more flattened distribution, with a scale height of only one or two thousand pc.

The division of the globular clusters into two families is not particularly convincing if only the spatial distributions are used. However, when metallicity information is added, the division becomes obvious. The distribution of metallicities is clearly bimodal, with the halo globular clusters peaking at $[Fe/H] = -1.5$, and the thick disk globular clusters peaking at $[Fe/H] = -0.5$. The widths of the two metallicity distributions are substantial, but small enough that there is only modest overlap. A metallicity of about $[Fe/H] = -0.8$ is a pretty good choice for dividing the two families. If the reader is not familiar with the figures in Zinn's paper, then he or she is urged to look at them carefully.

The problem with using the globular clusters to probe the shape and kinematics of the halo and thick disk is that there simply aren't enough clusters. For example, to determine how the velocity dispersion changes with height above the disk. it will take a sample with many more than 150 members. This is the primary reason for turning to the field stars.

## 3. Proper Motion Surveys

The classical approach to finding those rare high-velocity halo stars which just happen to be passing through the solar neighborhood (within a few hundred pc of the Sun) is to study stars with high proper motions. In recent years there has been grand progress in the observational study of large samples of stars drawn from proper-motion surveys (*e.g.* see Sandage & Fouts 1987, Carney & Latham 1987). Careful photometry allows reasonably accurate distances to be derived for each star, including the effects of reddening. Radial velocity measurements can then be combined with the observed proper motions to get a space motion and Galactic orbit for each star. To complete the picture, metallicities are needed. These can be estimated from photometry, but metallicities measured spectroscopically are preferable.

As an example of the kind of information produced by the proper-motion surveys, Figure 1 shows a plot of the velocity $v_{rot}$ (in the direction of the Galactic rotation) *versus* the metallicity [m/H] for 1152 stars from the Carney-Latham survey. The Halo population is pretty obvious in this diagram, with a median metallicity of about [m/H] = -1.6 and a mean $v_{rot}$ of 0 km s$^{-1}$ , but notice how there is a large dispersion in both the metallicity and $v_{rot}$ for the halo population, with a tail extending down to the metallicity of [m/H] = 0.0 and $v_{rot}$ of 225 km s$^{-1}$ expected for the thin disk. Indeed, there is a heavy concentration of points right where we expect the thin disk, but again with a considerable dispersion in both axes. As you inspect Figure 1, can you see a distinct third population, with median metallicity [m/H] = -0.6 and median $v_{rot}$ of about 175 km s$^{-1}$ ? Or, do the points in this region of the figure look more like a continuous extension of the thin disk to you? It is not yet clear which of these two interpretations should be preferred.

The problem with using surveys of nearby stars to probe the relation between kinematics and metallicity for the disk and halo populations is that the samples are biased. If a sample is selected from a proper-motion survey, then the metallicities should not be biased, but the kinematics will have selection effects which are very difficult to model. Conversely, if a sample is selected by metallicity, then the kinematics should be unbiased, but the metallicity selection effects will be hard to model. The solution to this dilemma is to observe halo and thick disk stars *in situ*, by looking above the Galactic plane, into regions where halo and thick disk stars predominate.

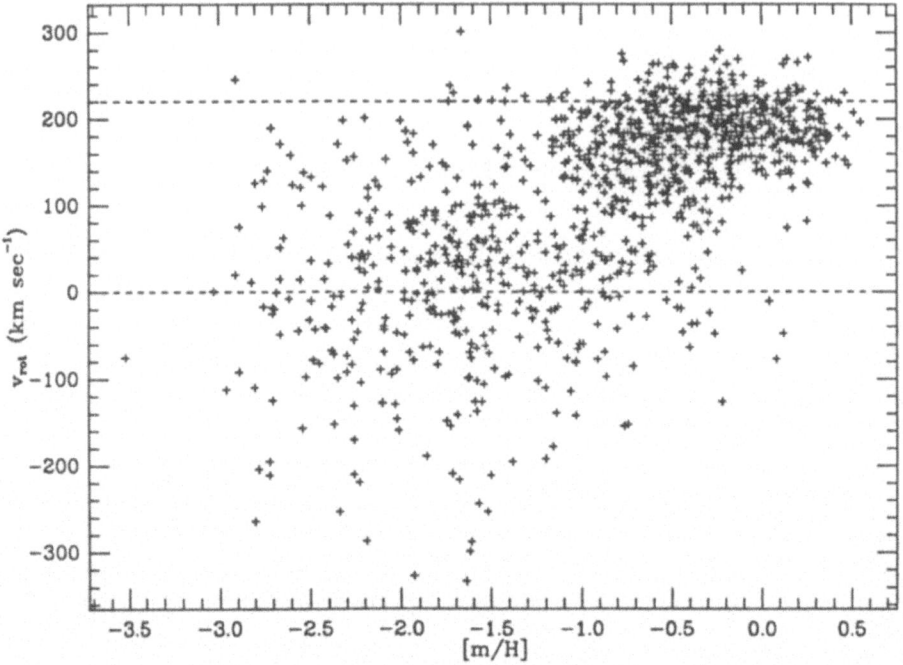

Fig. 1. The velocity, in the direction of Galactic rotation, $v_{rot}$, *versus* metallicity [m/H] for 1152 stars from the Carney-Latham proper-motion survey.

## 4. An *In Situ* Survey

For his thesis research at Harvard University, Ken Croswell carried out an *in situ* survey of 247 faint stars at the north Galactic pole (Croswell *et al.* 1991). Ken observed all the stars in nearly a square degree of Selected Area 57 (l = 66°, b = +86°) in the magnitude range $V$ = 13.31 to 17.88 and no redder than $B - V$ = 1.00. There should be no kinematical bias in this sample, while the metallicity biases introduced by the magnitude and color limits should be small.

Ken used the MMT to measure radial velocities good to about 10 km s$^{-1}$

TABLE I
**Observed $w$ velocity dispersion as a function of magnitude**

| $V$ range | Number of stars | $w$ velocity dispersion | predicted halo fraction |
|-----------|-----------------|-------------------------|-------------------------|
| 13-14 | 20 | $25 \pm 3$ | 0.164 |
| 14-15 | 53 | $46 \pm 4$ | 0.241 |
| 15-16 | 48 | $47 \pm 4$ | 0.362 |
| 16-17 | 54 | $80 \pm 7$ | 0.586 |
| 17-18 | 72 | $72 \pm 6$ | 0.863 |

for all 247 of the objects in his sample. For about half the stars he was able to obtain Strömgen photometry, which allowed him to derive luminosities, distances, and metallicities.

The point of going to faint stars in this survey was to rise out of the disk of the Galaxy to regions where the thick disk and halo stars predominate. That this goal was achieved is demonstrated by Table 1, where for each magnitude interval I list the number of stars, the observed velocity dispersion in km s$^{-1}$, and the fraction of the stars which should belong to the halo as predicted by the model of Bahcall and Soneira (1984). Notice how the observed velocity dispersion increases with magnitude, *i.e.* with distance above the disk. Even the nearest stars have a velocity dispersion which is larger than the value of roughly 20 km s$^{-1}$ which is appropriate for the $w$ velocity in the thin disk.

That Ken's survey was effective at eliminating most thin disk stars is also evident from Figure 2, which is the histogram of the $w$ velocity for all 247 stars. It is fairly clear that at least two populations are represented in Figure 2: the high-velocity tails are presumably members of the halo population, while the main lump with velocity dispersion of 30 or 40 km s$^{-1}$ in the center of the figure is presumably due mostly to thick disk stars. There is no obvious population of stars in Figure 2 with $w$ velocity dispersion of 20 km s$^{-1}$ , as would be expected for thin disk stars.

Unfortunately Ken was only able to measure Strömgen photometry for the stars in the brighter half of his sample. Since the contribution of the halo to the sample increases with distance, this means that he was unable to derive distances and metallicities for the more interesting half of his sample, the one that should include mostly halo stars. Nevertheless, the metallicity histogram for the 131 stars with Strömgen photometry, shown in Figure 3, is interesting. It is evident that the brighter half of Ken's sample consists mostly of thick disk stars, with metallicities in the range [Fe/H] = -0.3 to -1.0, although not all thin disk stars with metallicity near [Fe/H] = 0.0 were eliminated. Presumably the fainter half of Ken's sample would fill in the region near metallicity [Fe/H] = -1.5, where the metallicity distribution peaks for the halo globular clusters and for nearby halo dwarfs (Laird *et al.*

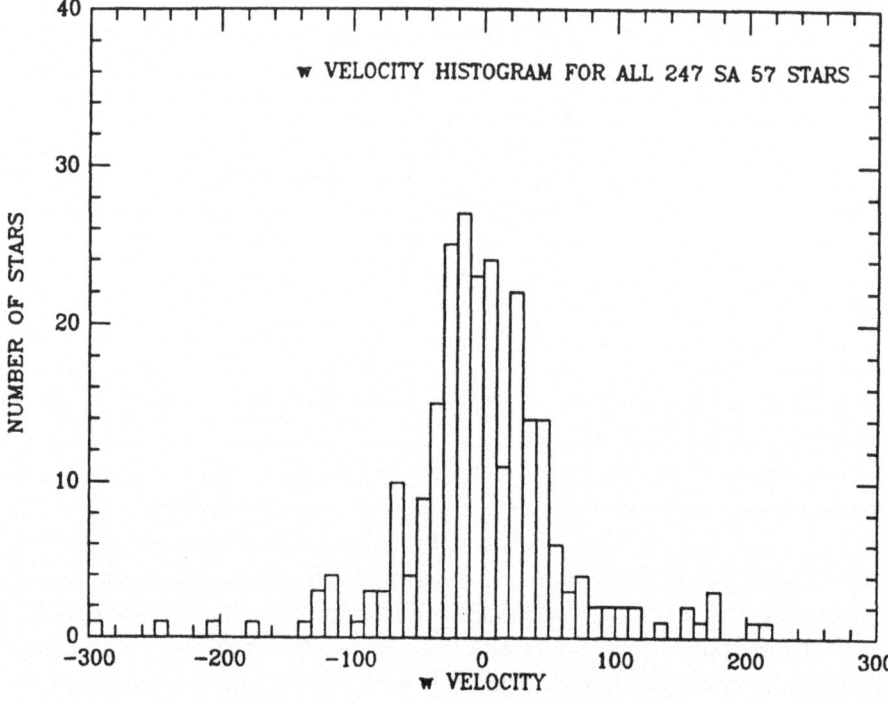

Fig. 2. The *w* velocity histogram for all 247 stars.

1988).

The plot of distance *versus* metallicity for the 131 stars with Strömgen photometry, Figure 4, shows how the more metal-poor halo stars are found at larger distances above the disk than their counterparts from the thick disk. Figure 4 leaves little doubt that there are at least two populations involved, but once again it is not obvious that a third discrete population is necessary. The thick disk may simply be the tail of the thin disk distribution. The separation of the sample into two populations may be less obvious in Figure 5, where *w* velocity is plotted *versus* metallicity. However, Figure 5 is useful for demonstrating that the velocity dispersion of the thick disk, *e.g.* for the stars with metallicity [Fe/H] more metal-rich than -1.0, is between

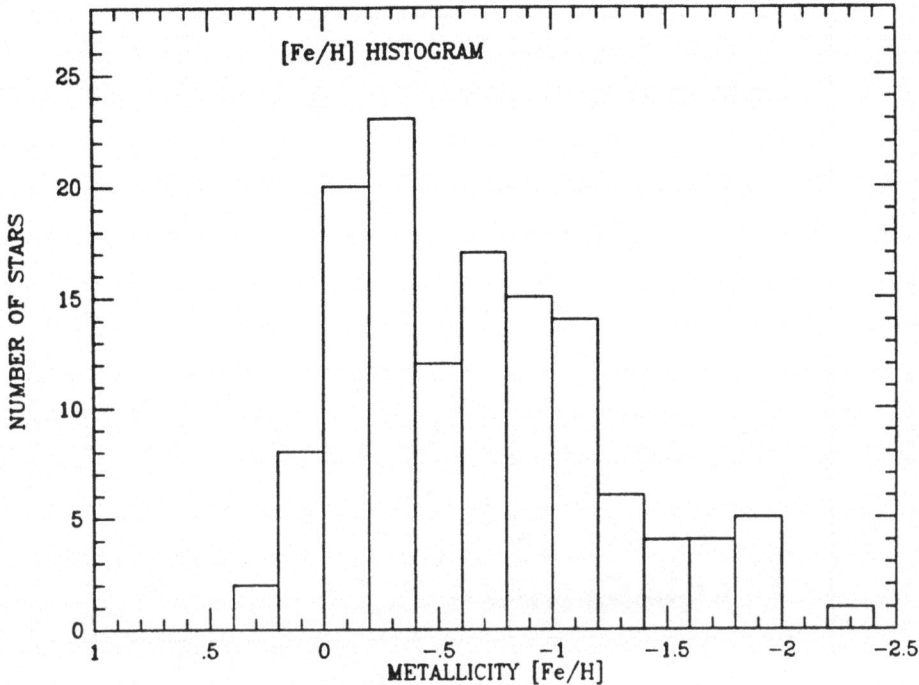

Fig. 3. The metallicity histogram for the 131 stars with Strömgen photometry.

30 and 35 $km\,s^{-1}$ .

Although Ken's results do not give definitive results for the distribution of stars in the disk and halo, they demonstrate that *in situ* surveys have great potential. This work needs to be extended by obtaining Strömgen photometry for the fainter half of the sample, and by extending the work to other selected areas. For example, surveys in directions orthogonal to SA 57, such as SA 51 ($l = 189°$, $b = +21°$) and SA 68 ($l = 111°$, $b = -46°$) could be used to probe the velocity ellipsoid of the halo along its other two axes.

Despite the formidable observational challenges involved, *in situ* surveys of the oldest populations of our Galaxy offer great promise for the study of

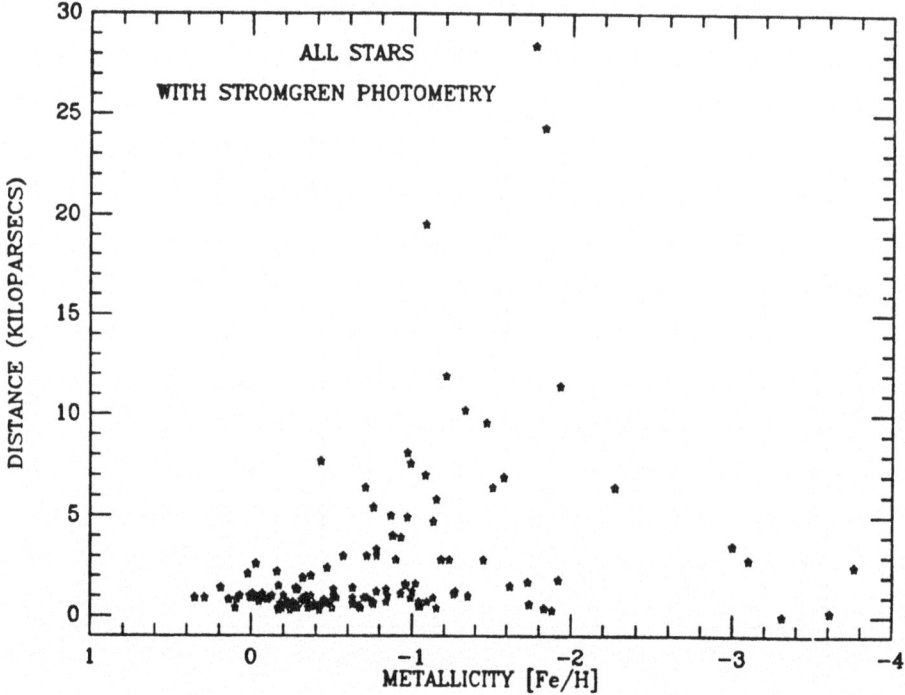

Fig. 4. Distance *versus* metallicity for the 131 stars with Strömgen photometry.

the structure and evolution of our Milky Way Galaxy.

## Acknowledgements

Much of this paper is based on Ken Croswell's thesis research, which had important contributions from Bruce Carney, Bill Schuster, and Luis Aguilar, and was supported in part by NSF, NASA, and SAO fellowships. My special thanks go to Bruce Carney for dragging me into Galactic research in the first place.

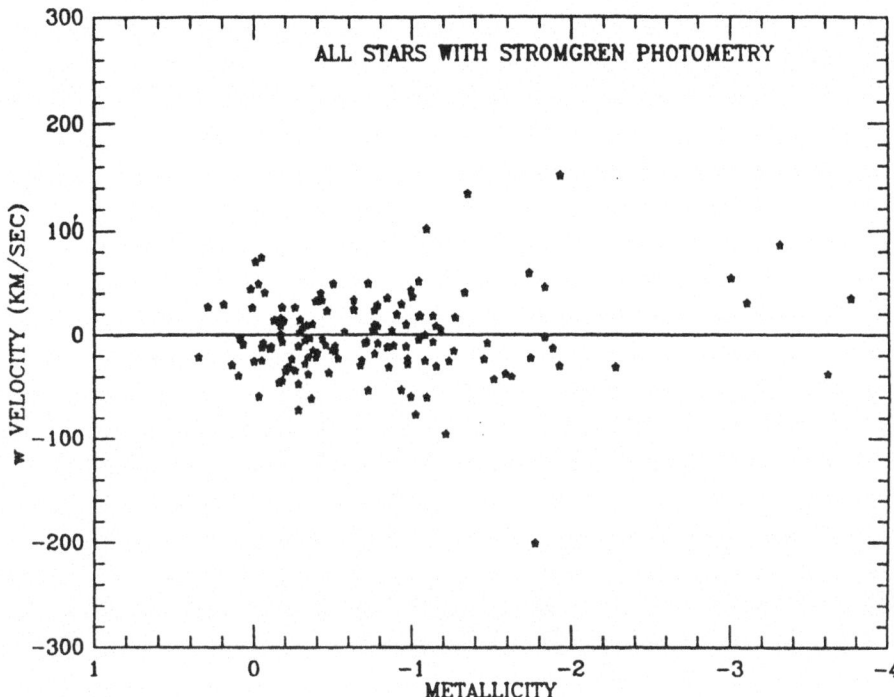

Fig. 5. $w$ velocity *versus* metallicity for the 131 stars with Strömgen photometry.

## References

Bahcall, J. N. & Soneira, R. M. 1984, *Ap. J. Suppl.*, **55**, 67.

Carney, B. & Latham, D. W. 1987, *A. J.*, **93**, 116.

Croswell, K., Latham, D. W., Carney, B. W., Schuster, W., & Aguilar, L. 1991, *A. J.*, **101**, 2078.

Laird, J. B., Rupen, M. P., Carney, B. W., & Latham, D. W. 1988, *A. J.*, **96**, 1908.

Norris, J. & Ryan, S. 1991, *Ap. J.*, **380**, 403.

Sandage, A. & Fouts, G. 1987, *A. J.*, **92**, 74.

Zinn, R. 1985, *Ap. J.*, **293**, 424.

The manufacturer's authorised representative in the EU is Springer
Nature Customer Service Centre GmbH, Europaplatz 3, 69115 Heidelberg,
Germany. If you have any concerns regarding our products, please
contact ProductSafety@springernature.com

Printed and bound by CPI Group (UK) Ltd, Croydon, CR0 4YY
29/04/2026
02099522-0010